GENETICS – RESEARCH AND ISSUES

NUTRIGENOMICS

APPLICATION TO THE DEVELOPMENT OF NUTRACEUTICALS AND COSMECEUTICALS

GENETICS – RESEARCH AND ISSUES

Additional books in this series can be found on Nova's website
under the Series tab.

Additional e-books in this series can be found on Nova's website
under the e-books tab.

NUTRITION AND DIET RESEARCH PROGRESS

Additional books in this series can be found on Nova's website
under the Series tab.

Additional e-books in this series can be found on Nova's website
under the e-books tab.

GENETICS – RESEARCH AND ISSUES

NUTRIGENOMICS

APPLICATION TO THE DEVELOPMENT OF NUTRACEUTICALS AND COSMECEUTICALS

M. T. RAVI SUBBIAH
EDITOR

NOVA BIOMEDICAL
New York

For permission to use material from this book please contact us:
Telephone 631-231-7269; Fax 631-231-8175
Web Site: http://www.novapublishers.com

Library of Congress Cataloging-in-Publication Data

ISBN: 978-1-62417-051-5

Library of Congress Control Number: 2012949232

Published by Nova Science Publishers, Inc. † New York

Contents

About the Editor

Dr. M.T. Ravi Subbiah, Ph.D, is Professor and Associate Director, Office of Laboratory Resources and Regulations in the Department of Internal Medicine at the University Of Cincinnati College Of Medicine, Cincinnati, Ohio. Dr. Subbiah received his initial education in India with B.Sc and M.Sc degrees from the University of Mysore and M.S. University of Baroda, respectively. He received his Ph.D degree in Biochemistry from the University of Toronto in Canada. He subsequently served as a Research Fellow, Assistant Professor and Consultant in the Division of Cardiovascular Diseases at the Mayo Clinic, Rochester, Minnesota. Dr. Subbiah joined the University of Cincinnati College of Medicine as an Associate Professor in 1978 and was appointed Professor in 1981. His research has focused upon the biochemical, endocrine and nutritional aspects of cholesterol metabolism and atherosclerosis. In recent years, his research has focused on therapeutic potential of estrogens in menopause and the development of safer estrogens and nutraceuticals based on nutrigenomics to improve women's health. Dr. Subbiah has also served as Vice President and Laboratory Director of a biomedical start-up company offering services in diagnostic genetic testing.

Dr. Subbiah is the author of more than 170 scientific publications, many chapters and has previously edited a book titled 'Atherosclerosis: A Pediatric Perspective'. He has served as a consultant for the TOKTEN program in India under the United Nations Development Fund. He has been a visiting scientist under the Burroughs-Welcome travel fellowship at Cambridge University and the University of Salford, England. He has been a member of many scientific organizations including Federation of American Societies for Experimental Biology, American Heart Association, and Society for Experimental Biology & Medicine, American Chemical Society and American Federation For Clinical Research.

Introduction

Nutrigenomics promises us the ability to tailor diet based on individual genetic make-up. Through the advances made in the Human Genome Project, single nucleotide polymorphisms in important genes have been identified and their clinical association with many metabolic imbalances has been documented. This has resulted in the creation of a nutrigenomic test panel that can be used to evaluated individual genomic differences in the clinical laboratory. The ability to interpret these nutrigenomic tests and perhaps recommend specific nutraceuticals and functional foods that can modify negative health effects of individual genetic profile is the hall mark of nutrigenomics. This way, nutrigenomics can play a vital role in personalized medicine.

While there has been a number of monographs dedicated to describe nutrigenomic concepts and their clinical associations very few books have incorporated authoritative examples of nutraceutical product development that could be of use in human subjects with selective metabolic problems (exm:cardiovascular problems).

This book also examines a neglected area related to the role of nutigenomics and nutraceuticals in skin health and dermagenetics. Dermagenetics is an off shoot of nutrigenetics which examines the role of selected genetic mutations in skin and determine what health supplements and nutraceuticals will offer the most benefit for individual skin. This area has developed into a multi-billion dollar industry and has been used by some companies to sell products to the public without solid evidence of its benefit when applied topically. No systematic reviews have analysed whether phytonutrients (so called cosmeceuticals) after direct topical application are as effective as nutraceuticals given orally in obtaining desired benefits. In three chapters we will examine the details on the role of nutrigenomics and epigenomics as effective modulators for cosmeceutical development, cutting edge formulation methods like nanotechnology for efficient release of these compounds from products useful in skin health.

We will cover newer approaches (membrane route) for developing nutraceuticals and also cite unexplored excellent sources to explore the nutrigenomic effects of its components and develop new nutriceuticals particularly related to fertility and sexual health.

In: Nutrigenomics
Editor: M. T. Ravi Subbiah

ISBN: 978-1-62417-051-5

Chapter I

Nutrigenomics and its Role in Personalized Medicine

M. T. Ravi Subbiah
Department of Internal Medicine, University of Cincinnati Medical Center,
Cincinnati, Ohio

Introduction

Nutrigenomics promises us the ability to formulate personalized diets optimal for our individual genotype to improve health [1]. The rapid growth in the nutrigenomic sciences was made possible by the marked advances in the Human Genome Project. This resulted in the identification of many single nucleotide polymorphisms (SNP's) in the key metabolic genes. Further linkage studies in the humans clearly established the association of selected gene mutations with specific metabolic disorders. This has given an opportunity for the food industry to formulate the dietary supplements that can negate the undesirable effects of specific gene mutations. Several genetic testing laboratories offer the tests (Nutrigenomic Test Panels) that can detect specific SNP's in the key metabolic genes [2]. Depending upon the genetic mutation identified in an individual, specific nutraceuticals and functional foods can be recommended to prevent any potential negative health effects [3]. This way, nutrigenomics can play a vital role in personalized nutrition and preventive medicine [2].

In the first section of this book (chapters 2-6), we describe the nutrigenomic concepts and its application to the development of the nutraceutical products. The primary focus is placed on the nutraceutical products that have the potential in managing the risk factors related to the cardiovascular diseases. This area has been extensively investigated and offers the most promise for its immediate application to human preventive nutrition.

The second section of this book (chapters 7-9) deals with the role of nutraceuticals and cosmeceuticals in skin health. An off-shoot of nutrigenomics called Dermagenetics, has developed in the recent years. Dermagenetics deals with a) the role of selected gene mutations in the skin health and b) the development of the health supplements, nutraceuticals and/or cosmeceuticals to improve the skin health based on individual genotype [4]. This area has been controversial with regard to the validity of the genes selected for the genetic testing and the efficacy of the cosmeceutical products. The consumer market for the cosmeceutical products has developed into a multi-billion dollar industry [5, 6]. The three chapters under this section examine: a) the role of dermagenetics, nutraceuticals and cosmeceuticals in the evaluation and maintenance of the skin health; b) the use of the nanotechnology based methods in cosmeceutical formulations [7] and c) the role of epigenetic factors in skin health and cosmeceutical development.

The third section of the book (chapters 10 and 11) explores the newer approaches and resources for the development of nutraceuticals. The chapters deal with a) the potential use of membrane routes for the cellular delivery of nutraceuticals and b) the unexplored plant sources suitable for the nutrigenomic studies and the development of new nutriceuticals related to fertility and sexual health.

The final chapter of this book (chapter 12) deals with the three issues that are critical for the successful application of nutritional genomics in personalized nutrition and also to protect the public from the commercial exploitations of genetic testing. These are: a) to set-up proper regulatory guidelines to evaluate the health claims of marketed products [8]; b) to make available in-depth nutrigenomic courses and training programs for the dietary and health professionals that can lead to the professional certification in this field; and c) to establish the state/federal regulatory and ethical guidelines [9] for those laboratories that offer the Direct to Consumer (DTC) genetic tests.

This book will enable the reader to get an in-depth understanding of the scope of nutrigenomics and its application to personalized nutrition and preventive health care.

Keywords: nutrigenomics, personalized medicine, single nucleotide polymorphisms, nutraceuticals, functional foods, dermagenetics, cosmeceuticals, nanotechnology, direct-to-consumer genetic tests

References

[1] Subbiah, MTR. Understanding the nutrigenomic definitions and concepts at the food-genome junction. *Omics J Integr Biol.* 2008; 12:1-7.

[2] Subbiah MTR. Nutrigenetics and nutraceuticals: the next wave riding on personalized medicine? *J Transl Res* 2007;149:55-61.

[3] Fenech M. Genome health, nutrigenomics and nutrigenetics: diagnosis and nutritional treatment of genome damage on an individual basis. *Food Chem Toxicol.* 2008;46:1365-1370.

[4] Draelos ZD. Cosmeceuticals: undefined, unclassified and unregulated. *Clin Dermatol.* 2009; 27:431-434.

[5] Subbiah MTR. Application of nutrigenomics in skin health. Nutraceutical or cosmeceutical? *J Clin Aesthet Dermatol.* 2010; 3:44-46.

[6] Gao XH, Zhang L, Wei H, Chen HD. Efficacy and safety of cosmeceuticals. *Clin Dermatol* 2008; 26:367-374.

[7] Boonme P, Junyaprasert VB, Suksawad N, Songkro S. Micro-emulsions and nanoemulsions: novel vehicles for whitening cosmeceuticals. *J Biomed Nanotechnol.* 2009; 5:373-383.

[8] Aggett P. The process of assessment of scientific support for claims on food. *Eur J Nutr.* 2009; 48(suppl): 523-526.

[9] Saukko PM, Reed M, Britten N, Hogarth S. Negotiating the boundary between medicine and consumer culture: online marketing of nutrigenomic tests. *Sci Soc Med.* 2011; 70:744-753.

Nutrigenomics and Nutraceutical Development

In: Nutrigenomics
Editor: M. T. Ravi Subbiah
ISBN: 978-1-62417-051-5

Chapter II

Definitions, Concepts and Application of Nutrigenomics in Health

M. T. Ravi Subbiah
Department of Internal Medicine, University of Cincinnati Medical Center, Cincinnati, Ohio, US

Abstract

The field of nutritional genomics promises us the ability to tailor diet based on individual genetic make-up utilizing the advances made in the Human Genome Project. A branch of nutigenomics called nutrigenetics provides us with a frame work for a) documentation of single nucleotide polymorphisms (SNP's) in candidate genes; b) explore their association with metabolic imbalances; and c) offers a panel of nutrigenomic tests that can be used in the clinical laboratory. Another branch called nutrigenomics examines a) the interaction of dietary components and botanicals with the genome and b) the ability of genetic variations to cause marked differences in individual response to dietary factors. Advances in nutritional genomics have led to the commercial development of nutraceuticals, dietary supplements and functional foods that are targeted to modify negative health effects of individual genetic profile. This chapter will also deal with an area that been largely ignored when evaluating the role nutraceuticals, i.e. dermagenetics and skin health. Dermagenetics (an off shoot of nutrigenetics) examines selected genetic mutations to determine what ingredients will offer benefit the individual skin health. Skin tissue is unique in that it can derive benefits of nutraceuticals given orally and also capable of assimilating phytonutrients (so called cosmeceuticals) after direct topical application. The latter aspect has developed into multi-billion dollar cosmeceutical industry. Commercial availability of skin care products that claim benefits in treating wrinkles, skin coloration and texture needs federal guidelines to validate the claims. In this chapter, we discuss the current status of concepts and definition of sub-fields related to nutritional genomics, need for guidelines for training of professionals in this area and explore its potential for the development of both nutraceuticals and cosmeceuticals.

Keywords: nutrigenomics, personalized medicine, single nucleotide polymorphisms, nutraceuticals, nutrigenetics, dermagenetics, cosmeceuticals, coronary artery disease, proteomics, transcriptomics, epigenetics, epigenomics, metabolomics, methylene tetrahydrofolate reductase, nitric oxide synthase, ω-3 fatty acids, ATP cassette transporters, phytosterols, phytostanols, interkeukins, skin health, ant-aging compounds, women's health, hormone replacement therapy, thrombosis

Introduction

Escalating health care costs are putting a heavy burden on the health care system in the United States. Long term care associated with the debilitating effects of chronic diseases like diabetes, coronary artery disease (CAD) and stroke contribute heavily towards these costs. In view of this, there is considerable interest in utilizing preventive medicine measures for early detection, management and treatment of the risk factors associated with chronic diseases. Development of chronic diseases in an individual are influenced both by environmental and genetic factors. Human Genome Project [1] has revolutionized medical research through documentation of individual variation in gene sequences, identification of single nucleotide polymorphisms (SNPs), their functional effects and thereby, providing an opportunity to elucidate their role in chronic diseases. In drug development, an area called pharmacogenomics [2,3], has made great strides in patient treatment by assessing the role of individual genetic make up in predicting drug efficacy and potential side effects (Personalized Medicine).

By knowing the effect of drugs in individuals with selected gene mutations in the early stages of drug development process, pharmaceutical companies will have better options to design drugs targeted to a given population groups, thereby reducing side effects. Similarly, diet and nutrition can have major influence on obesity and other risk factors associated with chronic diseases. We are all very well aware that 'one diet does not fit all' and recent investigations dealing with gene/nutrient interactions have provided a basis for individual differences in response to dietary components [4]. This new knowledge has given us the opportunity to tailor diet based on our individual genetic make-up (Personalized Nutrition) and thus, the beginnings of a scientific discipline called Nutritional Genomics!

Definition of Nutritional Genomic Sub-Disciplines

In the last decade, the scope of nutritional genomics has expanded considerably due to the advances in technology platforms [5] that go way beyond basic genetic analysis. The area of disease biomarker discovery is driving the field into proteomics, metabolomics, global transcriptomics etc. The rationale behind these explorations is to establish a system to translate data from thousands of gene mutations into a reasonable number of meaningful metabolites or biomarkers that can be easily measured in the laboratory.

Another important factor generating interest is the demonstration that hundreds of nutrients and botanicals interact with the genome altering gene expression and profile of proteins and other metabolites [6]. Some of the nutritional genomic terms and their definitions [6] based on their common usage in scientific publications are listed below:

1 Nutritional genomics: refers to the description of studies as to how the dietary factors affect the whole genome structure and function including a) the effect of individual dietary components on the genome and b) the differences among individuals in response to a given dietary factor.
2 Nutrigenomics: refers to the study of interaction of dietary factors with the genome and the resulting changes in gene expression, structure and function of proteins and other metabolites.
3 Nutrigenetics: examines individual differences in response to a give dietary factor based on their genetic make-up and identify their compatibility in terms of health or disease.
4 Proteomics: study of the differences in protein structure, profile and expression among individuals based on their genetic make-up, as well as the effect of dietary and other factors on these parameters [7].
5 Metabolomics: study of metabolic diversity (metabolite profile, metabolic functions and pathways) based on individual differences in genetic make-up, as well as their response to dietary and other factors on these parameters as discussed in a number of excellent reviews [8-10]).
6 Transcriptomics: deals with how altered expression of genetic variants contributes to complex diseases such as cancer, diabetes and heart disease. This provides researchers access to genome-wide differential expression data in order to obtain insight into metabolic pathways and molecular mechanisms that contributes to a given disease progression [11].
7 Nutraceuticals and Functional foods: Nutraceutical refers to any substance that is part of a food, which is capable of providing a health or medical benefit. This category includes vitamins, minerals and other diet supplements, specialized diets and genetically engineered foods. Functional foods on the other hand are defined as foods that can benefit human health beyond the effects of nutrients alone [12].
8 Epigenetics and Epigenomics: Epigenetics deals with changes (methylation, acetylation etc.) on the genome that are copied through generations and can modify gene expression without changes in primary DNA sequences [13]. The sum of genome-wide epigenetic patterns is called epigenomics. This phenotype expression can be modified by nutritional manipulation [13].

The vast number of nutrigenomic studies today are performed both in animals and in man and also in vitro model systems of peripheral mononuclear cells [14,15] followed by sophisticated gene expression and metabolomic analysis.

Genes of Nutrigenetic Interest

Although recent studies have identified numerous SNP's that are of interest from the nutrigenomic point of view, translation of these observations into nutrigenetic tests that can diagnose and provide dietary recommendations are limited in number [16]. One of the best studied nutrigenomic tests widely available in the laboratory are the two SNPs (C677T and A1298C mutations) in methylene tetrahydrofolate reductase (MTHFR) gene [17,18]. These MTHFR gene mutations cause reductions in 5-methyl tetrahydrofolate, (a co-factor for MTHFR enzyme activity) causing increased plasma homocysteine, a known risk factor for venous thromboembolic disease and neural tube defects [19,20]. Folic acid supplementation helps to overcome the negative health effect of these mutations in MTHFR gene [21,22].

During the last decade, a number of relevant mutations in lipoprotein metabolism (particularly with apo-E, apo A-1 and hepatic lipase genes) have been identified. These mutations seem to modify cardiovascular risk upon their interaction with dietary fat type and quantity [23-27].

For example, Subjects with apo-E 4 isoform respond negatively in terms of increased risk for coronary heart disease when exposed to a high fat diet, [24, 25]. In subjects with a G to A mutation in the promoter region of apo A-1 gene, greater amounts of polyunsaturated fat was required to raise plasma HDL levels when compared to normals. [26]. In recent years, mutations of interest between a) Factor VII R353Q polymorphism and fat intake causing increased blood cholesterol and triglycerides in females [28] and b) APOA5 and ω6 fatty acid polymorphism [29] has been identified.

Furthermore, genetic variants in fatty acid desaturases (catalyzing the metabolism of ω6 and ω3 fatty acids) are getting considerable attention in view of their importance in determining fatty acid profile and concentrations in phospholipid fraction of tissues and membranes [30-32]. New important discoveries in this field deal with the interaction of key genes related to obesity (exm: leptin, adiponectin) with fatty acids in subjects with metabolic syndrome [33,34]. Ferguson et al. [33] noted that subjects with metabolic syndrome carrying polymorphisms in adiponectin gene and its receptors showed greater insulin resistance and that lowering plasma saturated fatty acids (SFA) improved insulin resistance indicating personal dietary advice to decrease SFA consumption could improve insulin sensitivity.

Similarly, homozygotes for polymorphisms in leptin receptor gene demonstrated increased insulin resistance and these associations were abolished by a high (ω-3) and low (ω-6) polyunsaturated background [34].

Another important observation [35] deals with genes related to inflammation (a major risk factor for coronary artery disease) showing that polymorphisms in cytokines (IL-6, LTA and TNF-ά) increases risk factors for metabolic syndrome which is exacerbated at low polyunsaturated/saturated fatty acid diet. Polymorphisms in insulin receptor regulator (CAV2rs2270188) was shown to increase diabetes risk with higher dietary fat intake [36]. Nitric oxide synthase (NOS3) polymorphisms were associated with high plasma triglycerides which was responsive to polyunsaturated fatty acids [37]. Leptin and leptin receptor gene mutations have emerged as leading candidates towards predicting obesity [38-41].

Important nutrigenetic discoveries have also been made in genes related to oxidative stress (superoxide dismutase, glutathione peroxidase etc.), inflammation (IL-6, C-reactive protein) and sterol absorption (ATP cassette transporters) that are responsive to nutritional manipulation [6]. In the area of epigenomics, that deals with post translational modifications (methylation, acetylation etc.) and their effect on gene expression [13] is viewed as an important target site suitable for dietary and environmental manipulations. This area has been recently reviewed by Mathers [13] and Bayarsaihan [42].

Excellent reviews by Ordovas [15], Kornman [27], Lokiotnov [44], Allayee et al. [45] and Ferguson and Philpott [46] also critically review the scope of nutrigenetics and nutrigenomics in preventive health practice.

Nutrigenomic Advances and Its Potential for Nutraceutical Development

During the last decade a wealth of information on the effect of dietary factors, botanicals and herbs on the genome has been collected. This has allowed us to get to the actual mechanism of the beneficial effect of dietary factors on human health and some examples will be discussed here. A good example in this category is the cardioprotective effect of ω3 fatty acids [47]. Recent studies have shown that fish and fish oils (the main source of ω3 fatty acids in human diet) interfere at multiple stages in the atherosclerosis plaque development. This includes decreasing the generation of reactive oxygen species and attenuating expression of endothelium adhesion factors, and pro-inflammatory and pro-angiogenic enzymes through its effect on nuclear factor-kappa B system [47].

Similarly, the role of dietary isoflavones in vascular protection has been elucidated and shown to be due to its ability to activate antioxidant defense systems, decreasing the bioavailability of nitric oxide and modulating anti-inflammatory marker expression [48, 49]. The mechanism for the beneficial effect of whole grain cereals have been investigated [50] and shown to involve their multiple abilities (to serve as methyl donors, antioxidant protection due to phytic acid and lignin, content of b-complex vitamins, intestinal immunomodulatory effect, etc.). The potential role of dietary bioactive compounds in modulating apoptosis as a preemptive target in cancer has also been suggested [51].

Other studies have evaluated the effects of spices and their chemical components (exm: Curcuma longa) to control inflammation and oxidative stress [52]. Strong support for the nutrigenomics based benefits has also been obtained by large scale international studies. In a randomized dietary intervention study involving eight European countries (LIPOGENE), supplementation of ω3 polyunsaturated fatty acids along with a low fat diet, caused significant improvement in risk factors associated with metabolic syndrome [53]. In RIVAGE intervention study carried out in France [54], the observed responses to a Mediterranean diet clearly indicate the ability of the diet to interact with genome to favorably influence the cardiovascular risk. Nutrigenomics based studies in Spain [55] specifically showed the ability of polyphenol fraction of olive oil to down regulate expression of pro-atherogenic genes and improve cardiovascular risk factors.

In a major public health study [56] to determine the outcome of chronic diseases in response to a healthy diet and physical exercise (The French National Nutrition and Health Program), men showed significant reduction in risk of major chronic diseases (combination of cardiovascular disease and cancer or death). New study designs utilizing community based participatory research along with personalized medicine and nutrition [57] will be useful in sorting out relation of chronic disease risk to lifestyle factors as well evaluate translational research directives.

This ability of natural products to target specific metabolic sites and favorably manipulate critical steps in the progression of many chronic diseases has sparked a great interest to develop nutraceuticals, dietary supplements and functional foods. It is well known that chronic diseases like atherosclerosis, hypertension, type 2 diabetes, obesity etc. result from metabolic imbalances and they serve as major risk factors towards coronary heart disease [58-60]. Oxidative, inflammatory and thrombogenic mechanisms also contribute to the development of atherosclerosis [61-64]. Nutrients play an important role in influencing factors responsible for this metabolic imbalance and nutritional genomics will determine what (and how) nutrients will produce the desired impact on metabolic balance as influenced by individual genetic make up (personalized nutrition).

One of the classical examples to illustrate the nutrigenetic/nutraceutical concept is the factors influencing plasma cholesterol levels, a major risk factor for CAD. It has been known for nearly 50 years that response to dietary cholesterol among individuals varies dramatically. In fact, in the earlier literature subjects were generally classified as 'hyper' and 'hypo' responders to dietary cholesterol. In the last decade researchers have identified the role of ABC cassette transporters in cholesterol absorption [65, 66]. In subjects carrying polymorphisms (C1950G) in ABCG5 gene for example, increased absorption and response to dietary cholesterol have been noted [66]. In individuals carrying this mutation, including a dietary therapy (exm: plant sterol and/or its esters) to inhibit cholesterol absorption will be a good nutraceutical approach. Phytosterols have long been known to inhibit cholesterol absorption [67].

Recently, an ester consisting of a C-29 saturated plant sterol (sitostanol) esterified to mixed fatty acids, have been shown to an effective therapy to decrease plasma cholesterol. These esters, currently sold in the form of margarine Benecol in US grocery stores represent one of the new line of clinically tested products shown to decrease cholesterol absorption and lower plasma cholesterol [68].

Furthermore, these esters have also been incorporated into foods like milk and yogurt [69] providing choice and easy access to customers. Phytosterol esters represent a good example of opportunities available in nutraceutical development giving the public personal options to manage and improve their health.

Despite of the availability of drug (statin) therapy, there is a renewed public interest in using cost effective "dietary therapy" for life time maintenance of optimal plasma cholesterol levels. It is interesting to note the market for fortified and functional foods is projected to grow by at least 7% each year from 47 billion recorded in 2002 [70]. Using purified and clinically tested nutraceuticals will also eliminate health concerns about the risk of toxicity and potential interaction with other drugs [71,72] associated with the use of dietary supplements and herbs. It is hoped that a number of well tested nutraceuticals targeted to other sites of metabolic imbalance will find their way into the market place. Some of examples of potential nutraceutical/dietary supplement development, their target metabolic conditions and benefits are listed below:

1 Dietary supplement: containing resvertrol, antioxidant vitamins, green tea extract, poly unsaturated fatty acids and tomato extract to control inflammation and modulate oxidative and metabolic stress in overweight men [73].

2 Soybean product: to modify inflammation due to IL-1B, IL-2, IL-4, IL-8 and IL-10 polymorphism in relation to gastric cancer risk [74].
3 Avacado/soybean unsaponifiables: to inhibit nitric oxide synthase and matrix metalloproteinase 13 in experimental osteoarthritis [75].
4 Sorgum bran extracts: to decrease inflammation and hyaluronidase activity potentially useful in osteoarthritis [76].
5 Rho iso-alpha acids from hops: to inhibit GSK-3/NF kappa B pathway and decrease inflammation in collagen induced arthritis [77].
6 Short chain fructo-oligosaccharides: To modulate intestinal immunomodulatory and lipid lowering through peroxisome proliferator-activated receptor alpha and farnesoid receptor [78].

Application of Nutrigenomics to Skin Health: Special Considerations

The skin tissue is unique because it can derive benefits from orally administered nutraceuticals as well from topically applied phytonutrients called cosmeceuticals incorporated into skin creams [79,80]. Previous work has clearly shown that oral nutraceuticals can improve certain aspects of skin health [81,82]. This relationship between oral nutraceutical benefit and skin health has sparked a field called Dermagenetics (an off shoot of nutrigenetics).

In dermagenetics, tests for selected genetic mutations (that could potentially predict skin health) are done and interpreted to determine what nutraceuticals will offer benefit to the individual skin health. The common mutations tested are related to the enzymes involved in a) collagen breakdown, b) photo aging and free oxygen radical elimination, c) degradation of environmental pollutants and d) generation of pro-inflammatory molecules. These expensive dermagenetic tests are offered direct-to-consumers (without physician recommendation) by some companies, followed by advocating the use of specific nutraceuticals or skin creams containing selected cosmeceuticals.

The rationale, validity and benefit of these genetic tests are not very obvious. Some of the criticisms against advocating these tests are: a) due to the effect of post translational modifications and individual dietary make-up, one cannot predict that a given mutation will indeed display significant abnormalities at the blood and tissue levels; b) lack of substantiation of a given deficiency in an individual by any other laboratory tests; c) normal values for blood parameters related to any proposed parameters (exm: oxidative stress) are not available and d) therefore, advocating a nutraceutical or supplement (exm: antioxidant)is not justified.

In order for dermagenetics to be credible in offering benefit to consumers, other analytical tools like metabolomics [83], that actually evaluates metabolite profiles in blood and tissues under defined conditions is needed.

The use of cosmeceuticals in topical skin creams with claims of anti-aging, anti-wrinkle and collagen boosting agents has grown into a multi-billion dollar industry. There is considerable debate as to a) how much of these cosmeceuticals are absorbed from topical application and b) the extent of changes that actually occur in skin tissue [84].

Choi and Berson [85] concluded that the degree of benefits achieved by topical application might depend on a) formulation methods; b) the degree to which the active compounds can maintain integrity and c) the duration of their effects and d) factors affecting its release from the carrier vehicles. In fact, very little documentation of the actual changes in skin structure and cell biology has been examined after skin cream use.

Most studies measure benefit by subjective appearance of the skin [85]. Well controlled efficacy studies are needed to validate the claims of the skin care lotions to remove wrinkles, change skin coloration and texture [86, 87]. The application of nanotechnology [88, 89] to cosmeceutical development will certainly improve critical steps in the penetration of the active ingredients into the skin and may offer maximum benefits.

The Future of Nutrigenomics and Its Application in Personalized Medicine

Ongoing advances in high throughput technology and bioinformatics will help a) rapid identification of nutrigenetic gene mutations; b) establish their clinical validation in larger population studies; and c) continue to add new tests to the Nutrigenomic Test Panel (similar to Thrombogenic Panel [90], widely in use in clinical laboratories).

Large scale population studies [53-56] utilizing nutrigenomic approaches will provide evidence for the benefits of dietary factors (supplements, nutraceuticals) in selected metabolic imbalances. This will provide great opportunities to nutraceutical companies to create products in three distinct areas: a) dietary supplements like vitamins and herbal extracts; b) nutraceuticals incorporated into common (and ethnic) foods products in the form of juices, cereals, snacks, spreads etc.; c) whole functional foods (exm: margarine like Benecol). In the skin front, there will be a demand for functional foods that provide nutrients and those enriched with absorbable cosmeceuticals enhancing skin texture and function.

As new compounds preventing skin aging are identified there will also advancement in developing superior skin formulations for topical application that can be better absorbed. Nutraceuticals and cosmeceuticals suited toward genotype of individual and/or groups will have three major goals related to preventive medicine : a) to help alleviate a metabolic imbalance and fight against aging related changes (skin); b) decrease the amount of pharmaceutical use and c) ultimately to reduce individual drug and health care costs. One important point to remember is that the pharmacogenetics will also be applicable to drugs used in dermatology.

Therefore, combined with the advances in nutrigenomics, genomic expression profiling will become more routine and individualize both dermatologic therapy and nutrition [91, 92].

Finally, as the field (s) related to nutritional genomics expand their scope into nutraceutical, cosmetic and beauty products, it is essential that state/federal regulatory agencies a) set clear guidelines for diagnostic laboratories, especially for those offering genetic tests (including those laboratories testing for skin health) directly to consumers (without physician referrals); b) more strictly monitor health claims [93] of nutraceuticals and cosmeceuticals in skin products; c) work with the universities to train physicians, nutritionists and dietary professionals in nutrigenomics to meet the health challenges of this post-genomic era and d) deal with ethical issues in nutrigenomics [94].

References

[1] Venter, J. C., Adams, M. D., Myers, E. W., Peter, W., Mural, R. J., et al.: The sequence of human genome. *Science*. 2001; 291: 1304-135.

[2] Wang, L. Phramacogenomics: a systems approach. *Wiley Interdiscip. Rev. Syst. Biol. Med.* 2010;2: 3-22.

[3] Giacomino, K. M., Brett, C. M., Altman, R. B., Benowitz, N. L., Dolan, M. E., Flockhart, D. A., Johnson, J. A., et al.: Pharmacogenetic Research Network: From SNP discovery to clinical drug response. *Clin. Pharmacol. Ther.* 2007;81:328-345.

[4] Simopoulos, A. P. *Annu. Rev. Public Health*. 2010, 31: 53-68, 2010

[5] Garcia-Canas, V., Sino, C., Leon, C., Cifuentes, A. Advances in nutrigenomics research: Novel and future analytical approaches to investigate the biological activity of natural compounds and food functions. *J. Pharm. Biomed. Analy.* 2010; 51: 290-304.

[6] Subbiah, M. T. R. Understanding nutrigenomic definitions and concepts at the Food-genome junction. *Omics. J. Integr. Biol.* 2008; 12: 1-7.

[7] Kussmann, M., Affolter. Proteomics at the center of nutrigenomics: comprehensive molecular understanding of dietary effects. *Nutrition*. 2009; 25:1085-1093.

[8] German, J. B., Roberts, M. A., Watkins, S. M.: Personal metabolomics as a next generation nutritional assessment. *J. Nutr.* 2003; 133, 4260-4266.

[9] German, J. B., Roberts, M. A., Watkins, S. M.: Genomics and metabolomics as markers for the interaction of diet and health: lessons from lipids. *J. Nutr.* 2003;133: 2078S-2083S.

[10] Goodacre, R. Metabolomics of a superorganism. *J. Nutr.* 2007, 137: 259S-266S.

[11] Kato, H. Nutrigenomics: the cutting edge and Asian perspectives. *Asia. J. Clin. Nutr.* 2008; 17: 12-15.

[12] Ferguson, L. R. Nutrigenomic approaches to functional foods. *J. Amer. Diet. Assoc.* 2009; 109: 452-45.

[13] Mathers, J. C., Strathdee, G., Relton, C. L. Induction of epigenetic alterations by dietary and environmental factors. *Adv. Genet.* 2001; 71:3-39.

[14] Crujeiras, A. B., Parra, D., Milagro, F. I., Goyenechea, E., Larrante, E., Margareto, J., Martinez, J. A. Differential expression of oxidative stress and inflammation related genes in peripheral blood mononuclear cells in response to a low calorie diet: a nutrigenomic study. *Omics. J. Integrt. Biol.* 2008; 12: 251-261.

[15] Rudkowska, I., Raymond, C., Ponton, A., Jacques, H., Lavigne, C., Holub, B. J., Marette, A., Vohl, M. C. Validation of the use of peripheral blood mononuclear cells as surrogate model for skeletal muscle tissue in nutrigenomics tissue. *Omics. J. Integrt. Biol.* 2011; 15: 1-7.

[16] Subbiah, M. T. R. Nutrigenetics and nutraceuticals: the next wave riding on personalized medicine. *Transl. Res.* 2007;149: 55-6.

[17] Frosst, P., Blom, H. J., Milos, R., Goyette, R., Shepherd, C. A., Mathews, R. G., et al.: A candidate for genetic risk factor for vascular disease: common mutation in methylene tetrahydrofolate reductase. *Nature Genetics*. 1995; 10,111-113.

[18] Reyes-Engel, A., Munoz, E., Gaitan, M. J., Fabre, E., Gallo, M., Dieguez, J. L., et al.: Implication on the human fertility of the 677C to T and 1298A to C polymorphisms of the MTHFR gene: consequences of a possible genetic selection. *Mol. Human Reprod.* 2002; 8: 952-957.

[19] Hanson, H. Q., Aras, O., Yang, F., Tsai, M. Y.: C677T and A1298C polymorphisms of methylene tetrahydrofolate reductase gene: incidence and effect of combined genotypes on plasma fasting and post-methionine load homocysteine in vascular disease. *Clin. Chem.* 2001; 47: 661-666.

[20] Guieterez Revilla, J. I., Parez Hernandez, F., Calvo Martin, M. T., et al.: C677T and A1298C MTHFR polymorphisms in the etiology of neural tube defects in Spanish population. *Med. Clin.* (Barc.). 2003;120:441-445.

[21] Czeizel, A. E., Dudas, I.: Prevention of the first occurrence of neural tube defects by peri- conceptional vitamin supplementation. *N Engl. J. Med.* 1992; 327:1832-1835.

[22] Molloy, A. M., Scott, J. M.: Folates and prevention of disease. *Public Health Nutr.* 2001; 4: 601-609.

[23] Ordovas, J. M.: Genetic interactions with diet influence the risk of cardiovascular disease. *Am. J. Clin. Nutr.* 2006; 83 (suppl.) 443S-446S.

[24] Loctionov, A., Scollen, C., McKeowan, N., Bingham, S. A.: Gene-Nutrient interactions: dietary behavior associated with high coronary heart disease risk particularly affects serum LDL cholesterol in apolipoprotein E ε4 carrying free living individuals. *Br. J. Nutr.* 2000; 84: 885-890.

[25] Ordovas, J. M., Schaefer, E. J.: Genetic determinants of plasma lipid response to dietary intervention: the role of the ApoA1/C3/A4 gene cluster and APOE gene. *Br. J Nutr.* (suppl. 1). 2000; 83: S127-136.

[26] Ordovas, J. M., Corella, D., Cupples, L. A., Demissie, S., Kelleher, A., Coltel, O., et al.: Polyunsaturated fatty acids modulate the effects of the APOA1 G-A polymorphism on HDL cholesterol concentrations in a sex specific manner. The Framingham Study. *Amer. J. Clin. Nutr.* 2002; 75: 38-46.

[27] Ordovas, J. M., Corella, D., Demissie, S.: Dietary fat intake determines the effect of a common polymorphism in the hepatic lipase gene promoter on high density lipoprotein metabolism: evidence of a strong effect in this gene-nutrient interaction in Framingham Study. *Circulation.* 2002; 106: 2315-2321.

[28] Bowman, R., Joosen, A. M., Welch, A. A., Luben, R. N., Khaw, K. T., Wareham, N. J., Bingham, S. A. Blood lipids and fat intake: gene-nutrient interaction and risk of coronary heart disease with the factor VII R353Q polymorphism. *Eur. J. Clin. Nutr.* 2009; 63: 771-777.

[29] Park, J. Y., Paik, J. K., Kim, O. Y., Chae, J. S., Lee, J. H. Interactions between the APOA5-1131T>C and FEN110154G>T polymorphisms on ω6 polyunsaturated fatty acids in serum phospholipids and coronary artery disease. *J. Lipid Res.* 2010; 51: 3281-3288.

[30] Simopoulos, A. P. Genetic variants in the metabolism of omega-6 and omega-3 fatty acids: their role in the determination of nutritional requirements and chronic disease risk. *Exp. Biol. Med.* (Mayowood). 2010; 235: 785-795.

[31] Glaser, C., Heinrich, J., Koletzko, B. Role of EADS1 and FADS2 polymorphisms in polyunsaturated fatty acid metabolism. *Metabolism.* 2010, 59: 993-999.

[32] Lattka, E., Illig, T., Heinrich, J., Koletzko, B. *J. Nutrigenet Nutrigenomics.* 2009; 2:119-128.

[33] Ferguson, J. F., Phillips, C. M., Tierney, A. C., Perez-Martinez, P., Defoort, C., Helal, O., Lairon, D., et al. Gene-nutrient interactions in the metabolic syndrome: single nucleotide polymorphisms in ADIPOQ and ADIPOR1 interact with plasma saturated fatty acids to modulate insulin resistance. *Amer. J. Clin. Nutr.* 2009; 91: 794-801.

[34] Phillips, C. M., Goumidi, L., Bertrails, S., Field, M. R., Ordovas, J. M., Cupples, L. A., Defroot, C., et al. Leptin receptor polymorphisms interact with polyunsaturated fatty acids to augment risk of insulin resistance and metabolic syndrome in adults. *J. Nutr.* 2009; 140: 238-244.

[35] Phillips, C. M., Goumidi, L., Bertrails, S., Ferguson, J. F., Field, M. R., Kelly, E. R., Mehagen, J., et al. Additive effects of polymorphisms in the IL-6, LTA, TNF-alpha (genes) and plasma fatty acid level modulate risk for the metabolic syndrome and its components. *J. Clin. Endocrinol. Metab.* 2010; 95: 1386-1394.

[36] Fisher, E., Schreiber, S., Joost, H. G., Boeing, H., Doring, F. A two-step association study identifies CAV2rs2270188 single nucleotide polymorphism interaction with fat intake in type 2 diabetes risk. *J. Nutr.* 2011; 141: 177-181.

[37] Ferguson, J. F., Phillips, C. M., Mcmonagle, J., Perez-Martinez, P., Shaw, D. I., Lovegrove, J. A., Helal, O., et al. *Atherosclerosis.* 2010; 211:539-544.

[38] Li, W. D., Reed, D. R., Lee, J. H., Xu, W., Kilker, R. L., Sodom, B. R., Price, R. A.: Sequence variants in the 5'flanking region of the leptin gene are associated with obesity in women. *Ann. Human Genet.* 1999; 63: 227-234.

[39] Mammes, O., Betoulle, D., Aubert, R., Herpath, B., Siest, G., Fumeron, F.: Association of the G2548A polymorphism in the 5' region of the LEP gene with overweight. *Ann. Human Genet.* 2000;64: 391-394.

[40] Chagnon, Y. C., Wilmore, J. H., Borecki, I. B., Gagnon, J., Perusse, L., Chagnon, M., et al.: Associations between the leptin receptor gene and adiposity in middle aged Caucasian males from the HERITAGE family study. *J. Clin. Endo. Met.* 2000; 85: 29-34.

[41] Wauters, M., Mertens, I., Chagnon, M., Raniken, T., Considine, R. V., Chagnon, Y. C., et al.: Polymorphisms in the leptin receptor gene, body composition and fat distribution in overweight and obese women. *Int. J. Obesity.* 2001; 25:714-720.

[42] Bayarsaihan, D. Epigenetic mechanisms in inflammation. *J. Dent. Res.* 2011; 90: 9-17.

[43] Kornman, K. S.: Interleukin 1 genetics, inflammatory mechanisms and nutrigenetic opportunities to modulate diseases of aging. *Am. J. Clin. Nutr.* 2006; 83 (suppl.), 475S-483S.

[44] Loktionov, A.: Common gene polymorphisms and nutrition: emerging links with pathogenesis of mutifactorial chronic diseases. *J. Nutr. Biochem.* 2003; 14, 426-451.

[45] Allayee, H., Roth, N., Hodis, H. N. Polyunsaturated fatty acids and cardiovascular disease: implications for nutrigenetics. *J. Nutrigenet. Nutrigenomics.* 2009; 2:140-148.

[46] Ferguson, L. R., Philpott, M. Nutrition and mutagenesis. *Annu. Rev. Nutr.* 2008; 28:313-329.

[47] Massaro, M., Scoditti, E., Carluccio, M. A., Campana, M. C., De Caterina, R. Omega-3 fatty acids, inflammation and angiogenesis: basic mechanisms behind the cardioprotective effects of fish and fish oils. *Cell Mol. Biol.* 2010; 56: 59-82.

[48] Slow, R. C., Mann, G. E. Dietary isoflavones and vascular protection: activation of cellular antioxidant defenses by SERMs or hormones? *Mol. Aspects. Med.* 2010;31: 468-477.

[49] Blay, M., Espinel, A. E., Delgado, M. A., Balges, I., Blade, C., Arola, L., Salvado, J. Isoflavone effect on gene expression profile and biomarkers of inflammation. *J. Pharm. Biomed. Anal.* 2010; 51:382-390.

[50] Fardet, A. New hypothesis for the health protective mechanisms of whole-grain cereals: what is beyond fibre? *Nutr. Res. Rev*. 2010.23: 65-134.

[51] Martin, K. R. Using nutrigenomics to evaluate apoptosis as a target in cancer prevention. *Curr. Cancer Drug Targets*. 2007:7: 438-446.

[52] Zhou, H., Beevers, C. S., Juang, S. *Curr. Drug Targets*. 2011, 12:332-347.

[53] Tierney, A. C., Mcmonagle, J., Shaw, D. I., Gulseth, H. L., Saris, W. H., Paniaqua, J. A., Golabek-Lesszczynska, I., et al. Effects of dietary fat modification on insulin sensitivity and other risk factors of the metabolic syndrome-LIPOGENE: a European randomized dietary intervention study. *In. J. Obes*. (Lond.) 2010; 35:800-809.

[54] Lairon, D., Defoort, C., Martin, J. C., Amlot-Carlin, M. J., Gastaldi, M., Planells, R. Nutrigenetics: liks between genetic background and response to Mediterranean-type diets. *Public Health Nutr.* 2009; 12:1601-1606.

[55] Konstantinidou, V., Covas, M. I., Munoz-Aguayo, D., Khymenets, O., de la Torre, R., Saez, G., Tormos Mdel, C., et al. In vivo nutrigenomic effects of virgin olive oil polyphenols within the frame of the Mediterranean diet: a randomized controlled trial. *FASEB J.* 2010; 24:2546-2557.

[56] Estaquio, C., Castebon, K., Keese-Guyot, E., Bertrais, S., Deschamps, V., Dauchet, L., Peneau, S., et al. The French National Nutrition and health Program score is associated with nutritional status and risk of major chronic diseases. *J. Nutr.* 2008; 138:946-953.

[57] McCabe-Sellers, B., Lovera, D., Nuss, H., Wise, C., Ning, B., Teitel, C., Clark, B. S., et al. Personalizing nutrigenomics research through community based participatory research and omics technologies. *OMICS J. Integr. Biol.* 2008; 12: 263-272.

[58] Packard, C. J.: A triumvirate of targets in the prevention and treatment of paradigm for cardiovascular disease. *Atheroscler.* 2006; (Suppl.) 7: 21-29.

[59] Atilla, K., Vasan, R. S.: Pre-hypertension and risk of cardiovascular disease. *Expert Rev. Cardiovasc. Ther.* 2006; 4:111-117.

[60] Rosenberg, D. E., Jabbour, S. A., Goldstein, B. J.: Insulin resistance, diabetes and cardiovascular risk: approaches to treatment. *Diabetes Obes. Metab.* 2005;7: 642-653.

[61] Stocker, R., Keaney, J. F. jr: Role of oxidative modifications in atherosclerosis. *Physiol. Rev.* 2004; 84: 1381-1478.

[62] Tedgui, A., Mallat, Z.: Cytokines in atherosclerosis: pathogenic and regulatory pathways. *Physiol. Rev*. 2006.86; 515-581.

[63] Mach, F.: Inflammation is a crucial feature of atherosclerosis and a potential target to reduce cardiovascular events. *Handb. Exp. Pharmacol*. 2005; 170, 697-722.

[64] Voles-Gonzalez, J. F., Fuster, V., Badimon, J. J.: Links between inflammation and thrombogenecity in atherosclerosis. *Curr. Mol. Med*. 2006; 6: 489-499.

[65] Weggemans, R. M., Zock, P. L., Tai, E. S., Ordovas, J. M., Molhuisen, H. O., Katan, M. B.: ATP binding cassette G5 C1950G polymorphism may affect blood cholesterol concentrations in humans. *Clin. Genet.* 2002; 62, 226-229.

[66] Herron, K. L., McCrane, M. M., Waters, D., Lofgren, I. E., Clerk, R. M., Ordovas, J. M., Fernandez, M. L.: The ABCG5 polymorphism contributes to individual responses to dietary cholesterol and carotenoids in eggs. *J. Nutr.* 2006; 136,1161-1165.
[67] Subbiah, M. T. R. Dietary plant sterols: current status in human and sterol metabolism. *Amer. J. Clin. Nutr.* 1973; 26: 219-223.
[68] Thompson, G. R., Grundy, S. M.: History and development of plant sterol and stanol esters for cholesterol lowering properties. *Amer. J. Cardiol.* 2005; 96 (suppl. 1):3-9.
[69] Noakes, M., Clifton, P. F., Doornbos, A. M., Trautwein, E. A.: Plant sterol-ester enriched milk and yoghurt effectively reduce serum cholesterol in modestly hypercholesterolemic subjects. *Eur. J. Nutr.* 2005;44: 214-222.
[70] Sloan, A. E.: The top 10 functional food trends: The next generation. *Food Tech.* 2002; 56: 32-57.
[71] Cronin, J. R.: Dietary supplements: safety, certification, labeling and efficacy. *Alternat. Compliment Ther.* 2002; 8,141-148.
[72] Chan, M.: Herbal medicines: what work and how to use them safely. The Ohio Family Physician (fall). 2003; 20-23.
[73] Bakker, G. C. M., van Erk, M. J., Pellies, L., Wopereis, S., Ribingh, C. M., Cnubben, N. H. P. An anti-inflammatory dietary mix modulates inflammation and oxidative and metabolic stress in overweight men: a nutrigenomics approach. *Amer. J. Clin. Nutr.* 2010; 91:1044-1059.
[74] Ko, K. P., Park, S. K., Cho, L. Y., Gwack, J., Yang, J. J., Shin, A., Kim, C. S., et al. Soybean product intake modifies the association between interleukin-10 genetic polymorphisms and gastric cancer risk. *J. Nutr.* 2009; 139:1005-1012.
[75] Boileau, C., Martel-Pelletier, J., Caron, J., Msika, P., Guillou, G. B., Baudouin, C., Pelletier, J. P. Protective effects of total fraction of avocado/soybean unsaponifiables on the structural changes in experimental osteoarthritis: inhibition of nitric oxide synthase amd matrix metalloproteinase-13. *Arthritis. Res. Ther.* 2009;69:1179-1184.
[76] Bralley, E., Greenspan, P., Hargrove, J. L., Hartle, D. K. Inhibition of hyaluronidase activity by select sorghum brans. *J. Med. Food.* 2008; 11: 307-312.
[77] Konda, V. R., Desai, A., Darland, G., Bland, J. S., Tripp, M. L. Rho iso-alpha acids from hops inhibit GSK-3/ NF-kappa pathway and reduce inflammatory markers associated with bone and cartilage degradation. *J. Inflamm.* (London) 2009; 6:26.
[78] Fukasawa, T., Kamel, A., Watanabe, Y., Koga, J., Abe, K. Short chain fructooligosaccharide regulates hepatic peroxisome proliferators-activated receptor alpha and farnesoid X receptor target gene expression in rats. *J. Agric. Chem.* 2010; 58:7007-7012.
[79] Rattan, S. I. The science of healthy aging: genes, milieu and chance. *Ann. NY Acad. Sci.* 2007;1114:1-10.
[80] Bergstrom, K. G. Beyond tretinoin: cosmeceuticals for aging skin. *J. Drugs Dermatol.* 2009; 8:674-677.
[81] Udompataikul, M., Sripiroj, P., Palungwachira, P. An oral nutraceutical containing antioxidants, minerals and glycosaminoglycans improves skin roughness and fine wrinkles. *Int. J. Cosmet. Sci.* 2009; 31: 427-435.
[82] Izumi, T., Saito, M., Obata, A., Aril, M., Yamaguchi, H., Matsuyama, A. Oral intake of soy isoflavone aglycone improves the aging skin in adult women. *J. Nutr. Sci. Vitaminol.*, (Tokyo). 2007; 53:57-62.

[83] Xiao, J. F., Zhou, B., Ressom, H. W. Metabolite identification and quantitation in LC-MS/MS based metabolomics. *Trends Analyt. Chem.* 2012; 32: 1-14.

[84] Choi, C. M., Berson, D. S. Cosmeceuticals. *Semin. Cutan. Med. Surg.* 2006;25:163-168.

[85] Draelos, Z. D. Cosmeceuticals: undefined, unclassified and unregulated. *Clin. Dermatol.* 2009; 27: 431-434.

[86] Gao, X. H., Zhang, L., Wei, H., Chen, H. D. Efficacy and safety of innovative cosmeceuticals. *Clin. Dermatol.* 2008; 26:367-374.

[87] Bruce, S. Cosmeceuticals for the attenuation of extrinsic and intrinsic dermal aging. *J. Drugs Dermatol.* 2008; 7 (Suppl. 2):s17-22.

[88] Guglielmini, G. Nanostructured novel carrier for topical application. *Clin. Dermatol.* 2008;26: 341-346.

[89] Booneme, P., Junyaprasert, V. B., Suksawad, N., Songkro, S. Microemulsions and nanoemulsions: novel vehicles for whitening cosmeceuticals. *J. Biomed. Nanotechnol.* 2009; 5:373-383.

[90] Patnaik, M., Dlott, J. S., Fontaine, R. N., Subbiah, M. T. R., et al. Detection of genomic polymorphisms associated with venous thrombosis using invader biplex assay. *J. Mol. Diagn.* 2004; 6:137-144.

[91] Rizzo, A. E., Maibach, H. I. Personalizing dermatology: The future of genomic expression profiling in individualized dermatologic therapy. *J. Dermatolog. Treat.* 2012;23:161-167 (.

[92] Subbiah, M. T. R. Application of nutrigenomics in skin health: Nutraceutical or cosmeceutical? *J. Clin. Aesth. Dermatol.* 2010; 3:44-46.

[93] Aggett, P. The process for the assessment of scientific support for claims on foods. *Eur. J. Nutr.* 2009; 48 (suppl.): 523-526.

[94] Korthals, M. Coevolution of nutrigenomics and society: ethical considerations. *Amer. J. Clin. Nutr.* 2011;94: 2025s-2029s.

In: Nutrigenomics
Editor: M. T. Ravi Subbiah
ISBN: 978-1-62417-051-5

Chapter III

Nutrigenomic Effects of Olive Oil on Cardiovascular Risk

Montserrat Fitó,[1,2] Sandra Martín-Peláez,[1] Olga Castañer[1] and Valentini Konstantinidou[3]
[1]Cardiovascular Risk and Nutrition Research Group (CARIN), Research in Inflammatory and Cardiovascular Disorders (RICAD), Barcelona, Spain
[2]CIBER de Fisiopatología de la Obesidad y Nutrición, Instituto de Salud Carlos III, Spain
[3]Hellenic Health Foundation, Athens, Greece

Abstract

Adherence to traditional Mediterranean diet, with olive oil as its main source of fat, has been associated with a reduced risk of overall mortality, cardiovascular mortality, cancer incidence and mortality, and incidence of neurodegenerative diseases. The benefits associated with the Mediterranean diet and olive oil phenolic compounds consumption on cardiovascular risk can be mediated through changes in the expression of atherosclerosis-related genes. At present, limited data exists on the *in vivo* effect of the olive oil and the Mediterranean diet on human gene expression. In this chapter, we have attempted to review studies relating to the *in vitro* and *in vivo* nutrigenomic effects of olive oil and their phenolic components and evaluate its significance within the context of a healthy diet. The intervention with the traditional Mediterranean diet supplemented with virgin olive oil improved the expression of key genes involved in vascular inflammation, foam cell formation and thrombosis. Olive oil is able to modify *in vivo* gene expression in a protective fashion for proteins participating in the cellular mechanisms dealing with oxidative stress resistance, inflammation, lipid metabolism and other atherosclerosis-related traits/pathways such as insulin resistance. The phenolic compounds of olive oil have also been shown to influence the expression of pro-inflammatory and/or oxidative stress genes *in vivo* studies in healthy or individuals with metabolic syndrome. While peripheral blood mononuclear cells are usually used as models to explore changes of gene expression *in vivo* in humans, *in vitro* studies have reported data on the effect of phenolic compounds and other minor components of olive oil on atherosclerosis related gene expression in a variety of cell types (endothelial cells, mononuclear cells, adipocytes, hepatocytes etc). Intervention using a healthy diet with olive oil or its components can thus actively control the expression of pro-atherothrombotic genes and reduce cardiovascular risk in both healthy and cardiovascular high-risk individuals.

Abbreviations

ABCA1	ATP-binding cassette, sub-family A (ABC1), member 1
ADAM17	A disintegrin and metalloproteinase domain 17 TNFα, converting enzyme
ADRB2	Adrenergic beta-2-receptor
ALOX5AP	Arachidonate 5-lipoxygenase-activating protein
ARHGAP15	Rho-GTPase activating protein15
CCL2/MCP1	Chemokine (C-C motif) ligand 2
CD36	Thrombospondin receptor
COX-2	Cytochrome c oxidase subunit II
CPT-1	Carnitine palmitoyltransferase 1A
CV	Cardiovascular
CVD	Cardiovascular disease
DCLRE1C	DNA cross-link repair 1C
HDL	High density lipoprotein
ICAM-1	Intracellular cell adhesion molecule-1
IFNγ	Interferon gamma
IL10	Interleukin 10
IL7R	Interleukin 7-receptor
LDL	Low density lipoprotein
LRP1	Lipoprotein receptor-related protein 1
MD	Mediterranean diet
MMP-9	Matrix metalloproteinase-9
MUFA	Monounsaturated fatty acid
NFκB	Nuclear transcription factor kappa beta
OGT	O-UDP-N-acetylglucosamine (polypeptide-Nacetylglucosaminyl transferase
OO	Olive oil
PBMNCs	Peripheral Blood Mononuclear Cells
POLK	Polymerase (DNA directed) κ
PPARa	Peroxisome proliferator-activated receptor alpha
PPARγ	Peroxisome proliferator-activated receptor gamma
PPARBP	Peroxisome proliferator-activated receptor binding protein
PUFA	Polyunsaturated fatty acids
SFA	Saturated fatty acids
MD	Traditional Mediterranean diet
TNFα	Tumor necrosis factor alpha

USP48	Ubiquitin-specific protease 48
VCAM-1	Vascular cell adhesion molecule-1
VOO	Virgin olive oil

Keywords: nutrigenomics, mediterranian diet, cardiovascular risk, olive oil, atherosclerosis, pro-inflammatory genes, oxidative stress related genes, insulin sensitivity genes, cell surface expression genes, metabolic syndrome, olive oil phenolic compounds, monounsaturated fatty acids, polyunsaturated fatty acids, saturated fatty acids

1. Mediterranean Diet and Cardiovascular Health

Adherence to traditional Mediterranean diet (MD), with olive oil (OO) as its main source of fat, has been associated with a reduced risk of overall mortality, cardiovascular (CV) mortality, cancer incidence and mortality and incidence of Parkinson's and Alzheimer's disease [1-3]. The most impressive benefits of this diet, however, are related to reductions in CV morbidity and mortality [4].

The Mediterranean diet is also characterized by a) high consumption of vegetables, legumes, fruit and cereals; b) regular, but moderate, wine intake; c) moderate consumption of fish; d) from low to moderate intake of dairy products; and e) low consumption of meat. Olive oil consumption assures an increase in monounsaturated fatty acids (MUFA) intake, without a significant rise in saturated fatty acids (SFA) and guarantees an appropriate intake of polyunsaturated fatty acids (PUFA). The minor components of OO are classified into the unsaponifiable fraction and the soluble fraction. The former is defined as the solvent-extracted fraction after the saponification of the oil, whereas the latter includes the phenolic compounds [5].

2. Molecular Mechanisms of the Beneficial Effect of Olive Oil Ingestion

Olive oil components can exert healthy effects by a number of mechanisms. On one hand, exogenous antioxidant compounds, incorporated into the organism through diet, can exert their beneficial effects through antioxidant chemical reactions. Antioxidants and other cell redox state modulating enzyme systems, act as the first-line defence against free radicals not only in all cellular compartments, but also in extracellular space [6]. In the case of MUFA, it must be pointed out that it is less prone to oxidation [7].

On the other hand, the mechanism of action of nutrients in humans is strongly related to their capacity to modulate gene expression. Nutrigenomics is the study of molecular relationships between nutritional stimuli and the response of the genes [8]. From a nutrigenomic point of view, nutrients act as dietary signals, which detected by a cellular sensor influence gene and protein expression and, subsequently, metabolite production [9]. In addition, nutrigenomics aims to identify genes that can influence the risk of diet-related diseases on a genome-wide scale.

3. Nutrigenomic Effects of Olive Oil Consumption on Cardiovascular Risk

Olive oil is able to modify gene expression coding in a protective mode for proteins participating in the cellular mechanisms involved in oxidative stress resistance [10], inflammation [11], lipid metabolism [12] and other atherosclerosis-related traits/pathways [13-15] such as insulin resistance [16]. The analysis of gene expression changes in peripheral blood mononuclear cells (PBMNCs) of healthy individuals (n=11), after an acute (50mL) and sustained (25mL/day) virgin olive oil (VOO) consumption, showed that interferon gamma (*IFNG*) expression decreased in microarray analysis [15, 17]. The expression of the following genes related to: lipid metabolism (ATP-binding cassette, sub-family A member 1 (*ABCA1*), PPAR binding protein *(PPARBP)*), inflammation (interleukin 10 *(IL10)*, ubiquitin specific peptidase 48 *(USP48)*, and ADAM metallopeptidase domain 17 *(ADAM17)*), oxidative stress (lipoic acid synthetase *(LIAS)*, O-linked N-acetylglucosamine (GlcNAc) transferase *(OGT)*) and DNA repair (DNA polymerase kappa *(POLK)*, DNA cross-link repair 1C *(DCLRE1C)*) were up-regulated. Therefore, these genes could be targets related to atherosclerosis, specifically modulated by OO ingestion within the context of varying dietary patterns (such as acute, sustained and moderate consumption of OO) and OO consumption within the context of a MD.

It is also important to point out that in real-life conditions VOO is never consumed alone but as part of a whole dietary pattern. Llorente-Cortés *et al.* [18] compared the short-term effects of two MDs (one supplemented with VOO and the other with nuts) versus a low-fat control diet on the expression of pro-atherogenic genes in monocytes isolated from blood before and three months after the dietary intervention in 49 high CV-risk participants (23 men, 26 women), aged 55-80 years. Intervention using the MD supplemented with VOO prevented an increase in cytochrome c oxidase subunit-II *(COX-2)* and low density lipoprotein receptor-related protein 1 *(LRP1)* and reduced the expression of monocyte chemoattractant protein 1 *(MCP1)*, key genes involved in vascular inflammation, foam cell formation and thrombosis.

3.1. Genes Related with Inflammation Processes

Oxidative stress, an imbalance between free radical formation and antioxidant status, induces inflammation via pathways that generate inflammatory mediators such as adhesion molecules and pro-inflammatory cytokines. Oxidative stress and inflammation are intertwined processes which induce vascular damage including impairing endothelial function and increasing arterial rigidity [19], as well as angiogenesis, lipid peroxidation and platelet activation [20].

It has been demonstrated that the intake of a VOO-based breakfast is able to suppress *in vivo* expression of several pro-inflammatory genes in PBMNCs [21, 22]. Jiménez-Gómez *et al.* [22] evaluated the chronic effects of dietary fat on the postprandial expression of pro-inflammatory genes in the PBMNCs of 20 healthy men who followed three diets with different fat contents for 4 weeks. On the morning of the sampling, three different breakfasts (either with butter-, OO-, or walnut-base) was given to the volunteers. The butter-based breakfast elicited a higher increase in tumor necrosis factor alpha *(TNFA)* mRNA expression than the OO- or walnut-based ones. Butter- and walnut-enriched meals, but not OO-based ones, also elicited a nuclear transcription factor kappa beta *(NFKB)* postprandial activation in the PBMNCs of healthy volunteers [21]. These authors concluded that the consumption of an OO-enriched meal did not activate *NFKB* in monocytes as compared to the butter- and walnut-enriched ones and moreover, that this effect could enhance the cardioprotective effect of OO-enriched diets.

In another 3 month intervention study [23], with a high CV risk Mediterranean population (314 men and 407 women), it was studied the effect of both polymorphisms (the -765G>C in *COX2* and the -174G>C in interleukin-6 *(IL6)*) on inflammation markers (serum concentrations of IL6, C-reactive protein (CRP), intercellular adhesion molecule 1 (ICAM-1) and vascular cell adhesion molecule-1 (VCAM)). Although both polymorphisms were associated with inflammation, consuming a MD (either supplemented with VOO or nuts) reduced the concentration of inflammation markers regardless of these polymorphisms.

Adipose tissue macrophages are the source of inflammatory pathways and obesity, a well known CV risk factor [24]. Raff *et al.* tested the differences between diets with conjugated linoleic acids versus OO on gene expression changes in adipose tissue from 81 healthy postmenopausal women for 16 weeks [25]. The mRNA expression of glucose transporter type 4 *(GLUT4)*, leptin *(LEP)* and lipoprotein lipase *(LPL)* was lower and *TNFA* was higher in the linoleic acid-group versus the OO-control group. Similarly, 20 abdominally overweight subjects participated in a parallel, controlled-feeding trial for 8 weeks. A MUFA-diet led to a greater anti-inflammatory gene expression profile whereas an SFA diet resulted in a pro-inflammatory "obesity-linked" profile. Moreover, a decrease in serum low density lipoprotein cholesterol (LDL-cholesterol) and an increase in plasma and adipose tissue oleic acid content were also observed. These results indicate a prevention of adipose tissue inflammation when dietary SFA are replaced by MUFA [26].

The phenolic compounds of OO have also shown to influence the expression of pro-inflammatory genes [27] *in vivo* in humans. Following a double-blind, randomized, crossover design, two VOO-based breakfasts with high (398ppm) and low (70ppm) content of phenolic compounds were administered to 20 patients suffering from metabolic syndrome. Expression changes in genes involved in inflammatory processes mediated by transcription factor *NFKB*, activator protein 1 *(AP-1)*, cytokines, mitogen-activated protein kinases (*MAPKs*), or arachidonic acid pathways were identified. The intake of VOO-based breakfast, which is rich in phenol compounds, was able to suppress the expression of several pro-inflammatory genes, thereby switching activity of PBMNCs to a less deleterious inflammatory profile.

Konstantinidou *et al.*, focusing on the effects of phenolic compounds, published a randomized, parallel, controlled trial in healthy volunteers (n=90) aged 20 to 50 years who consumed MD either with VOO (328mg/kg polyphenol content) or washed OO (55mg/kg polyphenol content) [11]. Both MD consumptions, particularly the one with the higher amount of polyphenols, decreased plasma oxidative and inflammatory status and the expression of genes related with both inflammation (*IFNG*, Rho GTPase activating protein 15 *(ARHGAP15)*, interleukin 7 receptor *(IL7R)*) and oxidative stress (adrenoceptor beta 2, surface *(ADRB2)*, *(POLK)*) in PBMNCs. These findings indicate a significant role of OO polyphenols in the down-regulation of proatherogenic genes within the context of a MD [11]. In the case of *IFNG*, there was an agreement between transcriptomic and systemic analyses i.e. the decrease in *IFNG* was observed both at phenotypic and gene expression levels. *IFNG* is considered to be a key inflammatory mediator for inducing *IL6*, a prime regulator of CRP synthesis in the liver [28]. The observed down-regulation of the *IFNG* expression in PBMNCs is in agreement with that observed in previous results after a single dose of VOO [17]. The protein encoded by *IL7R* is a receptor for interleukin-7 which has been associated with inflammatory processes [29]. A recent study has shown an up-regulation of stress-response genes, such as *IL7R* and *POLK*, after carbon ion irradiation in murine tumour models [30]. *POLK* is a DNA repair gene which copies undamaged DNA templates and it is unique among human Y-family DNA polymerases [31]. Somatic DNA mutations, promoted by DNA oxidation, are considered to be a crucial step in carcinogenesis as well as being involved in atherosclerosis [32, 33]. In agreement with the DNA oxidation analyses in the EUROLIVE study [34], the down-regulation of the *POLK* gene expression was associated with the MD intervention, but was not evident with the OO polyphenol content [11]. All these data suggest a protective role for MUFA, or other minor components of OO, on DNA oxidation and damage within the frame of the MD. The *ADRB2* gene was also down-regulated after 3 months of MD intervention, particularly in the MD+VOO intervention group [11]. At postprandial state, however, an increase in the *ADRB2* gene expression in human PBMNCs after 50 mL of VOO ingestion was observed [16], which could be considered a compensatory mechanism for the increased postprandial oxidative stress. *ADRB2,* a major lipolytic receptor in human fat cells, has been related to a protective effect against oxidative stress through the promotion of glutathione synthesis [35]. Intervention with a healthy diet can thus actively modulate the expression of pro-atherothrombotic genes in both healthy and CV high-risk individuals.

In vitro studies have reported data on the effect that phenolic compounds and other minor components of OO, exert on inflammation-related gene expression. A report in human cell cultures showed that OO polyphenols inhibit p38/CREB phosphorylation resulting in a downstream reduction of *COX2* expression [36]. Martínez-González *et al.* observed that oleanolic acid contributed to vascular homeostasis by inducing prostacyclin release in a *COX2* dependent manner in human coronary smooth muscle cells [37].

Most intracellular reactive oxygen species are generated as a product of oxidative phosphorylation in mitochondria [38]. Hao *et al.* proposed that hydroxytyrosol could have a protective effect on the development of CVD by a stimulation of mitochondrial biogenesis leading to the enhancement of mitochondrial function and cellular defence systems [39]. They investigated the effects of hydroxytyrosol in 3T3-L1 adipocytes. It was found that hydroxytyrosol stimulated the transcriptional activation and protein expression of peroxisome proliferator-activated receptor gamma, coactivator 1 alpha *(PPARGC1A)* (central factor for mitochondrial biogenesis) and its downstream targets; these included nuclear respiration factors 1 and 2 and mitochondrial transcription factor A which led to an increase in mitochondrial DNA (mtDNA). Treatment with hydroxytyrosol resulted in an enhancement of mitochondrial function, including an increase in activity and protein expression of Mitochondrial Complexes I, II, III and V, increased oxygen consumption and a decrease in free fatty acid contents in the adipocytes. The mechanistic study of the *PPARGC1A* activation signalling pathway demonstrated that hydroxytyrosol is an activator of 5'AMP-activated protein kinase and also up-regulates gene expression of peroxisome proliferator-activated receptor *alpha (PPARA)*, carnitine palmitoyltransferase 1A (*CPT1)* and peroxisome proliferator-activated receptor *gamma (PPARG)* [39]. Reactive oxygen species (ROS) and matrix metalloproteinase 9 (MMP9) have been associated with CV disorders. Nakbi *et al.*

[40] studied the capacity of VOO tyrosol and hydroxytyrosol in impairing superoxide production and *MMP9* expressions in monocyte cells (THP-1) that differentiated into adherent macrophages taken as a model of human macrophages involved with atheroma formation. They observed a reduction of the expression of *MMP9* in conditioned medium derived from tyrosol and hydroxytyrosol treated cells.

3.2. Cell Surface Expression Genes

When the endothelial cells are exposed to inflammatory stimuli they express vascular cell adhesion molecules and mediate the adhesion and migration of monocytes to the sub-endothelial space, an initial step in atherosclerosis. There is evidence concerning the protective role of OO against vascular risk through the down-regulation of adhesion molecules involved in early atherogenesis. Carluccio *et al.* reported that oleuropein reduced monocytoid cell adhesion to stimulated endothelium, as well as *VCAM1* mRNA and protein, which is an essential early step in atherogenesis [41].

Dell'Agli *et al.* studied the role of the phenolic extract from extra VOO (oleuropein aglycone, hydroxytyrosol) and homovanillyl alcohol on cell surface and mRNA expression in human umbilical vascular endothelial cells, due to the fact that ICAM1, VCAM1 and E-selectin are crucial for endothelial activation [42]. The phenolic extract from extra VOO strongly reduced cell surface expression of *ICAM1* and *VCAM1*, linked to a reduction in mRNA levels. Oleuropein aglycone and hydroxityrosol were the main components responsible for these effects. Homovanillyl alcohol inhibited cell surface expression of all the adhesion molecules, whereas the effect on mRNA expression was weaker.

3.3. Genes Related with Insulin Sensitivity

Insulin resistance and compensatory hyperinsulinemia are involved in the development of dyslipidemia, hypertension, impaired fibrinolysis and other abnormalities that collectively contribute to an increased risk of CVD [43, 44]. The ability of an individual to cope with a fatty meal, which is largely a postprandial phenomenon, may be a key factor in the development of CVD [45].

Post-prandial changes in healthy volunteers were reported in the expression of the following target-genes mechanistically relevant for insulin sensitivity (*ADAM17*, *ABRB2*, arachidonate 5-lipoxygenase-activating protein (*ALOX5AP)*, cluster determinant 36 *(CD36)*, *LIAS*, *PPARBP*, *OGT*,) [16]. The postprandial time course of glucose, insulin and the expression of *CD36* were similar: they showed a peak at 1h after VOO ingestion, which returned to baseline values at 6 h. The opposite pattern (negative quadratic trend) was observed for *OGT ALOX5AP* gene expression changes. The observed increase in *OGT* expression at 6h after VOO ingestion could be linked to feedback mechanisms that develop to attenuate the signalling of sustained insulin action [46]. At this time, the observed upregulation of *CD36*, a fatty acid transporter [47], could be related to the postprandial increase in plasma fatty acids after VOO ingestion, as a related mechanism to the fatty acid uptake [48]. The increase in *CD36* could also be associated with the satiety response after OO ingestion. The activation of the small intestinal lipid messenger oleoylethanolamide, enabled by *CD36*-mediated uptake of dietary oleic acid, serves as a molecular sensor, linking fat ingestion to satiety [49]. Induction of *CD36* in response to these diverse stimuli, including oxidised LDL, is dependent on the activation of *PPARG* [50].

In agreement with this, a direct relationship was observed between *CD36* and *PPARBP*) gene expressions at 1h and 6h after OO ingestion. In addition, an oxidative stress status was present at 6h postprandial, as reflected by the increase in the oxidative damage to lipids and LDL [16]. The linear increase observed in triglycerides and oxidised LDL from baseline to 6h, at the postprandial state, was similar to that observed for *ADAM17*, *ADRB2*, *LIAS* and *PPARBP* gene expression changes. The protein encoded by the *LIAS* gene belongs to the lipoic acid synthetase family. Lipoic acid is a powerful antioxidant that can activate peroxisome proliferator activated receptors (*PPARA* and *PPARG*) [51]. It could, therefore, be hypothesised that the *LIAS* up-regulation and that of *PPARBP*, a *PPARG* co-activator, might be one of the feedback mechanisms for counteracting the postprandial oxidative stress involved in the development of insulin resistance [52]. Activation of *PPARG* by *PPARBP* may also increase insulin sensitivity by down-regulating the expression of *TNFA* [53]. *TNFA* is the major negative regulator of the insulin receptor pathway. *TNFA* is regulated at post-transcriptional level by the *TNFA* converting enzyme (*ADAM17*) [54]. Thus, the *ADRB2* upregulation observed at 6h after VOO ingestion could be promoted by the postprandial insulin peak [16], which would account for a delayed adrenergic signal for both adipocyte differentiation and oxidative stress protection [35].

Similar changes were observed after a real-life dose of OO, as it is consumed daily in some Mediterranean areas [55]. These results provided a first insight that the protective effect, related to VOO consumption at postprandial state observed in markers for CVD, could be mediated through gene expression changes. In previous works [15], all the above-mentioned genes with the exception of *ALOX5AP* and *ABRB2* had also increased their expression after 3 weeks of sustained and moderate (25mL per day) OO consumption. The strongest responders were *ADAM17*, *OGT*, *LIAS* and *PPARBP*. *ADAM17* and *OGT* were two of the highest up-regulated genes in both microarrays performed at a post-prandial state six hours after an acute dose and after a long-term moderate and regular consumption of VOO [15, 17]. After an acute 50mL VOO ingestion a postprandial oxidative stress response occurs [56] which is involved in the development of insulin resistance [57]. It could, therefore, be hypothesized that several feedback mechanisms, which involve up or down regulation of those genes, exist to counteract this stress. If so, that might explain the differences observed in the expression of several genes after acute and regular or moderate long-term consumption of VOO.

3.4. Hepatic Genetic Expression

Estévez-González *et al.* carried out a study for six weeks using skim milk or OO-enriched milk on 36 prepubertal children with mild hypercholesterolemia [58]. They examined the changes in serum lipids including its relationship to modulation of lipid levels by the Taq 1B polymorphism in the cholesteryl-ester transfer protein gene. It was observed that the consumption of skim milk enriched with OO increased high density lipoprotein cholesterol (HDL-cholesterol) and apolipoprotein-A1 (APOA1) levels, an effect that was more intense in carriers of the B1B1 genotype. The strong association of the cholesteryl-ester transfer protein gene single nucleotide polymorphisms (SNPs) (-4,502C>T promoter and TaqIB which are in partial linkage disequilibrium) and HDL-cholesterol was not statistically modified by diet or by the other environmental factors, in 4,210 high CV-risk individuals [59].

The impact of OO and its related compounds on hepatic genetic expression has been more extensively studied in animal models. Martinez-Beamonte *et al.* characterized the hepatic gene expression response to a bolus of 5ml of extra-VOO in male rats in order to understand the role of the liver as a gatekeeper that regulates postprandial lipemia [60]. Hepatic alpha-2-macroglobulin *(A2m), sodium-dependent citrate transporter (Slc13a5)* and neuronal regeneration related protein *(Nrep)* mRNA expressions were found to be significantly changed. They showed the highest significant associations with postprandial plasma triglycerides and a lack of association with basal triglyceridemia. These results highlight new gene regulation in liver by postprandial triglyceridemia.

Osada *et al.* studied the effects of the amount of dietary fat and saturation together with cholesterol on hepatic apolipoproteins in male rats [61]. The consumption of a diet supplemented with cholesterol and OO increased the levels of hepatic APOA1 compared to the control cholesterol group after 2-months of treatment. In a similar study in rats, Calleja *et al.* investigated the effects of the amount and saturation of dietary fat together with cholesterol on both hepatic *ApoA1* mRNA and plasma levels of this apolipoprotein [62]. Compared to the cholesterol diet, 40% and 10% OO diets induced a significant increase in the expression of *ApoA1* mRNA. A positive and significant correlation between plasma APOA1 concentration and its hepatic mRNA was observed when the amount of dietary OO was 40% (w/w). Guillen *et al.* tested in mice the hypothesis that the maslinic acid of OO dramatically influences hepatic gene expression [63]. Two OO, only differing in the presence of maslinic acid, were prepared (either 10% OO or 10% OO (w/w) enriched with maslinic acid). Hepatic gene expression was analyzed in apolipoprotein-E (*ApoE)* deficient mice with a C57BL/6J genetic background. Only cytochrome P450, family 2, subfamily b, polypeptide 9 *(Cyp2b9)*, cytochrome P450, family 2, subfamily b, polypeptide 13 *(Cyp2b13)* and D site albumin promoter binding protein *(Dbp)* expressions appeared to be significantly increased and macrophage receptor with collagenous structure (*Marco*) was significantly decreased in *ApoE* deficient mice receiving the diet enriched with maslinic acid. *Dbp* was up-regulated to some extent, depending on the genetic background of the mice and negatively associated with the expression of *Marco*, a gene strongly up-regulated by the absence of *ApoE.* It was concluded that maslinic acid in VOO is highly active in controlling hepatic gene expression and that there is a marked interaction between the response to maslinic acid and the presence of *ApoE.*

PBMNCs are being usually used to explore changes in gene expression *in vivo* in humans because they: 1) are critically involved in the atherosclerotic plaque formation; 2) had a role in the inflammatory pathways which lead to atherosclerosis; 3) are easily available from volunteers considering the feasibility of collection plus deontological reasons. Some discrepency between the transcriptomic and proteomic data could be due to different time-serial response, half-life of proteins, limitations of methodologies and/or different cellular lines involved in the expression of a gene. The progress in understanding the significance of nutrigenomic science is hindered by the fact that in human studies gene expression changes after an intervention with a diet-pattern or a single food (even more with real-life doses) are modest.

Probably the magnitude of nutrigenomic changes is smaller than that one caused by other types of manipulations (such as a pharmacological or pathologic processes). OO consumption elicited changes in the expression of genes related to CV risk-related processes such as insulin resistance, oxidative stress, and inflammation and lipid profile. The observed gene expression changes were in a protective mode to counteract these situations. The benefits associated with the MD and OO phenolic compound consumption, on CVD risk can be mediated through changes in the expression of atherosclerosis-related genes.

Acknowledgments

We thank Stephanie Lonsdale for help in editing the English text. The work of these authors was supported by grants from I.K.Y. Greek State Scholarship Foundation, CICYT (SAF 2004-08173-CO3-00), FIS-FEDER (PI041308), SNS Miguel Servet's (CP06/00100), Sara Borrell (CD10/00224) and, Rio Hortega (CM08/00054) contracts from Health Institute of Carlos III and, European Regional Development Found-ERDF. CIBER de Fisiopatología de la Obesidad y Nutrición (CIBEROBN) is an initiative of ISCIII (CB06/03/0028).

References

[1] Trichopoulou, A., Costacou, T., Bamia, C., Trichopoulos, D. Adherence to a Mediterranean diet and survival in a Greek population. *New England Journal of Medecine.* 2003; 348:2599-2608.

[2] De Lorgeril, M., Salen, P., Martin, J. L., et al. Mediterranean dietary pattern in a randomized trial: prolonged survival and possible reduced cancer rate. *Archives of Internal Medecine.* 1998; 158:1181-1187.

[3] Sofi, F., Cesari, F., Abbate, R., et al. Adherence to Mediterranean diet and health status: metaanalysis. *British Medical Journal.* 2008; 337:1344.

[4] Parikh, P., McDaniel, M. C., Ashen, M. D., et al. Diets and cardiovascular disease: an evidence-based assessment. *Journal of the American College of Cardiology.* 2005; 45:1379-1387.

[5] Tripoli, E., Giammanco, M., Tabacchi, G., et al. The phenolic compounds of olive oil: structure, biological activity and beneficial effects on human health. *Nutrition Research Reviews.* 2005; 18:98-112.

[6] Karihtala, P., Soini, Y. Reactive oxygen species and antioxidant mechanisms in human tissues and their relation to malignancies. *APMIS.* 2007;115:81-103.

[7] Perez-Jimenez, F., Alvarez de Cienfuegos, G., Badimon, L., et al. International conference on the healthy effect of virgin olive oil. *European Journal of Clinical Investigation.* 2005; 35:421-4.

[8] Chavez, A., Munoz, D. C. Nutrigenomics in public health nutrition: short-term perspectives. *European Journal of Clinical Nutrition.* 2003;57 Suppl. 1:S97-100.

[9] Muller, M., Kersten, S. Nutrigenomics: goals and strategies. *Nature Reviews Genetics.* 2003;4(4):315-322.

[10] Oliveras-López, M. J., Berná, G., Carneiro, E. M., *et al.* An extra virgin olive oil rich in polyphenolic compounds has antioxidant effect in OF1 mice. *Journal of Nutrition.* 2008;138:1074-1078.

[11] Konstantinidou, V., Covas, M. I., Muñoz-Aguayo, D., et al. *In vivo* nutrigenomic effects of virgin olive oil polyphenols within the frame of the Mediterranean diet: a randomized controlled trial. *FASEB J.* 2010;24:2546-57

[12] Eletto, D., Leone, A., Bifulco, M., Tecce, M. F. Effect of unsaturated fat intake from Mediterranean diet in rat liver mRNA expression profile: selective modulation of genes involved in lipid metabolism. *Nutrition, Metabolism and Cardiovascular Disease.* 2005;15:13-23.

[13] Deng, X., Elam, M. B., Wilcox, H. G., et al. Dietary olive oil and menhaden oil mitigate induction of lipogenesis in hyperinsulinemic corpulent JCR:LA-cp rats: microarray analysis of lipid-related gene expression. *Endocrinology.* 2004;145:5847-5861.

[14] Acin, S., Navarro, M. A., Perona, J. S., et al. Microarray analysis of hepatic genes differentially expressed in the presence of the unsaponifiable fraction of olive oil in apolipoprotein E-deficient mice. *British Journal of Nutrition.* 2007;97:628-638.

[15] Khymenets, O., Fitó, M., Covas, M. I., et al. Mononuclear cell transcriptome response after sustained virgin olive oil consumption in humans: an exploratory nutrigenomics study. *OMICS.* 2009;13:7-19.

[16] Konstantinidou, V., Khymenets, O., Covas, M. I., et al. Time course of changes in the expression of insulin sensitivity-related genes after an acute load of virgin olive oil. *OMICS.* 2009;13:431-8.

[17] Konstantinidou, V., Khymenets, O., Fito, M., et al. Characterization of human gene expression changes after olive oil ingestion: an exploratory approach. *Folia Biologica (Praha).* 2009;55:85-91.

[18] Llorente-Cortés, V., Estruch, R., Mena, M. P., et al. Effect of Mediterranean diet on the expression of pro-atherogenic genes in a population at high cardiovascular risk. *Atherosclerosis.* 2010;208:442-50.

[19] Lavi, S., Yang, E. H., Prasad, A., et al. The interaction between coronary endothelial dysfunction, local oxidative stress and endogenous nitric oxide in humans. *Hypertension.* 2008;51:127–133.

[20] Yung, L., Leung, F., Yao, X., et al. Reactive oxygen species in vascular wall. *Cardiovascular and hematological disorders drug targets.* 2006:6:1–19.

[21] Bellido, C., López-Miranda, J., Blanco-Colio, L. M., et al. Butter and walnuts, but not olive oil, elicit postprandial activation of nuclear transcription factor kappaB in peripheral blood mononuclear cells from healthy men. *American Journal of Clinical Nutrition.* 2004;80:1487-1491.

[22] Jiménez-Gómez, Y., López-Miranda, J., Blanco-Colio, L. M., et al. Olive oil and walnut breakfasts reduce the postprandial inflammatory response in mononuclear cells compared with a butter breakfast in healthy men. *Atherosclerosis.* 2009;204:e70-e76.

[23] Corella, D., González, J.I., Bulló, M., et al. Polymorphisms cyclooxygenase-2 -765G>C and interleukin-6 -174G>C are associated with serum inflammation markers in a high cardiovascular risk population and do not modify the response to a Mediterranean diet supplemented with virgin olive oil or nuts. *Journal of Nutrition.* 2009; 139:128-34.

[24] Weisberg, S. P., McCann, D., Desai, M., et al. Obesity is associated with macrophage accumulation in adipose tissue. *Journal of Clinical Investigation.* 2003; 112:1796-1808.
[25] Raff, M., Tholstrup, T., Toubro, S., et al. Conjugated linoleic acids reduce body fat in healthy postmenopausal women. *Journal of Nutrition.* 2009; 139:1347-1352.
[26] Van Dijk, S. J., Feskens, E. J., Bos, M. B., et al. A saturated fatty acid-rich diet induces an obesity-linked pro-inflammatory gene expression profile in adipose tissue of subjects at risk of metabolic syndrome. *American Journal of Clinical Nutrition.* 2009:90:1656-64. Epub. 2009 Oct. 14.
[27] Camargo, A., Ruano, J., Fernandez, J. M., et al. Gene expression changes in mononuclear cells in patients with metabolic syndrome after acute intake of phenol-rich virgin olive oil. *BMC Genomics.* 2010:11:253.
[28] Coppack, S. W. Pro-inflammatory cytokines and adipose tissue. *Proceedings of the Nutrition Society.* 2001. 60:349-356.
[29] Mobini, R., Andersson, B. A., Erjefalt, J., et al. A module-based analytical strategy to identify novel disease-associated genes shows an inhibitory role for interleukin 7 Receptor in allergic inflammation. *BMC systems biology.* 2009; 3:19.
[30] Imadome, K., Iwakawa, M., Nojiri, K., et al. Upregulation of stress-response genes with cell cycle arrest induced by carbon ion irradiation in multiple murine tumors models. *Cancer Biology and Therapy.* 2008; 7: 208-217.
[31] Kunkel, T. A., Pavlov, Y. I., Bebenek, K. Functions of human DNA polymerases eta, kappa and iota suggested by their properties, including fidelity with undamaged DNA templates. *DNA Repair (Amsterdam)* 2003; 2:135-149.
[32] Bogani, P., Galli, C., Villa, M., Visioli, F. Postprandial anti-inflammatory and antioxidant effects of extra virgin olive oil. *Atherosclerosis.* 2007; 190:181-186.
[33] Cooke, M. S., Evans, M. D., Dizdaroglu, M., Lunec, J. Oxidative DNA damage: mechanisms, mutation and disease. *FASEB Journal.* 2003; 17:1195-1214.
[34] Machowetz, A., Poulsen, H. E., Gruendel, S., et al. Effect of olive oils on biomarkers of oxidative DNA stress in Northern and Southern Europeans. *FASEB Journal* 2007; 21:45-52.
[35] Takahata, Y., Takarada, T., Iemata, M., et al. Functional expression of beta (2) adrenergic receptors responsible for protection against oxidative stress through promotion of glutathione synthesis after Nrf2 upregulation in undifferentiated mesenchymal C3H10T1/2 stem cells. *Journal Cell Physiology.* 2009; 218:268-275.
[36] Corona, G., Deiana, M., Incani, A., et al. Inhibition of p38/CREB phosphorylation and COX-2 expression by olive oil polyphenols underlies their anti-proliferative effects. *Biochemical and biophysical research communications.* 2007; 26; 362:606-11.
[37] Martínez-González, J., Rodríguez-Rodríguez, R., González-Díez, M. Oleanolic acid induces prostacyclin release in human vascular smooth muscle cells through a cyclooxygenase-2-dependent mechanism. *Journal of Nutrition.* 2008. 138:443-8
[38] Cadenas, E., Davies, K. J. Mitochondrial free radical generation, oxidative stress and aging. *Free Radical Biology and Medecine.* 2000:29,222–230.
[39] Hao, J., Shen, W., Yu, G., et al. Hydroxytyrosol promotes mitochondrial biogenesis and mitochondrial function in 3T3-L1 adipocytes. *Journal of Nutritional Biochemistry.* 2010; 21:634-44.

[40] Amel, N., Dabbou, S., Champion, S., et al. Modulation of the superoxide anion production and MMP-9 expression in PMA stimulated THP-1 cells by olive oil minor components: Tyrosol and hydroxytyrosol. *Food Research International.* 2011 (*in press).*
[41] Carluccio, M. A., Siculella, L., Ancora, M. A., et al. Olive oil and red wine antioxidant polyphenols inhibit endothelial activation: antiatherogenic properties of Mediterranean diet phytochemicals. *Arteriosclerosis Thrombosis and Vascular Biology.* 2003;23,622–629.
[42] Dell'Agli, M., Fagnani, R., Mitro, N., et al. Minor components of olive oil modulate proatherogenic adhesion molecules involved in endothelial activation. *Journal of Agricultural and Food Chemestry.* 2006; 3; 54:3259-64.
[43] Zavaroni, I., Bonini, L., Gasparini, P., et al. Hyperinsulinemia in a normal population as a predictor of non-insulin-dependent diabetes mellitus, hypertension and coronary heart disease: the Barilla factory revisited. *Metabolism.* 1999; 48:989-994.
[44] Steiner, G., Lewis, G. F. Hyperinsulinemia and triglyceride-rich lipoproteins. *Diabetes.*1996; 45 Suppl. 3:S24-S26.
[45] Laatsch, A., Merkel, M., Talmud, P. J., et al. Insulin stimulates hepatic low density lipoprotein receptor-related protein 1 (LRP1) to increase postprandial lipoprotein clearance. *Atherosclerosis.* 2009; 204:105- 111.
[46] Yang, X., Ongusaha, P. P., Miles, P. D., et al. Phosphoinositide signalling links O-GlcNAc transferase to insulin resistance. *Nature.* 2008; 451(7181):964-969.
[47] Corpeleijn, E., Pelsers, M. M., Soenen, S., et al. Insulin acutely upregulates protein expression of the fatty acid transporter CD36 in human skeletal muscle *in vivo. Journal of Physiology and Pharmacology.* 2008.59:77-83.
[48] Axelsen, M., Smith, U., Eriksson, J. W., et al. Postprandial hypertriglyceridemia and insulin resistance in normoglycemic first-degree relatives of patients with type 2 diabetes. *Annals of Internal Medicine.* 1999; 131: 27-31.
[49] Schwartz, G. J., Fu, J., Astarita, G., et al. The lipid messenger OEA links dietary fat intake to satiety. *Cell Metabolism.* 2008; 8: 281-288.
[50] Nicholson, A. C. Expression of CD36 in macrophages and atherosclerosis: the role of lipid regulation of PPARgamma signaling. *Trends in Cardiovascular Medecine.* 2004; 14:8-12.
[51] Pershadsingh, H. A. Alpha-lipoic acid: physiologic mechanisms and indications for the treatment of metabolic syndrome. *Expert Opinion in Investigational Drugs.* 2007;16:291-302.
[52] Giugliano, D., Ceriello, A., Esposito, K. Glucose metabolism and hyperglycemia. *American Journal of Clinical Nutrition.* 2008; 87:217S-222S.
[53] Itoh, H., Doi, K., Tanaka, T., et al. Hypertension and insulin resistance: role of peroxisome proliferator-activated receptor gamma. *Clinical and Experimental Pharmacology and Physiology.* 1999; 26:558-560.
[54] Serino, M., Menghini, R., Fiorentino, L., et al. Mice heterozygous for tumor necrosis factor-alpha converting enzyme are protected from obesity induced insulin resistance and diabetes. *Diabetes.* 2007; 56:2541-2546.
[55] Helsing, E. Traditional diets and disease patterns of the Mediterranean, circa 1960. *American Journal and Clinical Nutrition.* 1995; 61 (6 Suppl.):1329S-1337S.

[56] Fito, M., Gimeno, E., Covas, M. I., et al. Postprandial and short-term effects of dietary virgin olive oil on oxidant/antioxidant status. *Lipids.* 2002; 37:245-251.

[57] Giugliano, D., Ceriello, A., Esposito, K. Glucose metabolism and hyperglycemia. *American Journal and Clinical Nutrition.* 2008; 87:217S-222S.

[58] Estévez-González, M. D., Saavedra-Santana, P., López-Ríos, L., et al. HDL cholesterol levels in children with mild hypercholesterolemia: effect of consuming skim milk enriched with olive oil and modulation by the TAQ 1B polymorphism in the CETP gene. *Annals of Nutrition and Metabolism.* 2010; 56:288-93.

[59] Corella, D., González, J. I., Bulló, M., et al. Polymorphisms cyclooxygenase-2 -765G>C and interleukin-6 -174G>C are associated with serum inflammation markers in a high cardiovascular risk population and do not modify the response to a Mediterranean diet supplemented with virgin olive oil or nuts. *Journal of Nutrition.* 2009; 139:128-34.

[60] Martinez-Beamonte, R., Navarro, M. A., Guillen, N., et al. Postprandial transcriptome associated with virgin olive oil intake in rat liver. *Frontiers in Bioscence.* 2011; 1; 3:11-21.

[61] Osada, J., Fernández-Sánchez, A., Diaz-Morillo, J. L., et al. Differential effect of dietary fat saturation and cholesterol on hepatic apolipoprotein gene expression in rats. *Atherosclerosis.* 1994; 108:83-90.

[62] Calleja, L., Trallero, M. C., Carrizosa, C., et al. Effects of dietary fat amount and saturation on the regulation of hepatic mRNA and plasma apolipoprotein A-I in rats. *Atherosclerosis.* 2000; 152:69-78.

[63] Guillen, N., Acín, S., Surra, J. C., et al. Apolipoprotein E determines the hepatic transcriptional profile of dietary maslinic acid in mice. *Journal of Nutritional Biochemistry.* 2009; 20: 882-93.

In: Nutrigenomics
Editor: M. T. Ravi Subbiah
ISBN: 978-1-62417-051-5

Chapter IV

Dietary Mushrooms: An Excellent Source for Nutraceuticals Capable of Reducing Risk Factors for Cardiovascular Disease

W. J. Tucker,[1] S. S. Angadi[2] and K. R. Martin[1*]
[1]Nutrition Program, Healthy Lifestyles Research Center,
School of Nutrition and Health Promotion, Arizona State University
[2]Exercise and Wellness Program, Healthy Lifestyles Research Center,
School of Nutrition and Health Promotion, Arizona State University

Abstract

Cardiovascular disease (CVD) is the leading cause of morbidity and mortality in the developed world. Epidemiological evidence suggests that dietary intervention is an important and effective means of reducing CVD risk due to the plethora of beneficial bioactive agents present primarily in plant-based diets. Mushrooms such as shiitake (*Lentinus edodes*), maitake (*Grifola frondosa*) and oyster (*Pleurotus ostreatus*) in particular are rich in bioactive compounds. These compounds include eritadenine, lovastatin, beta-glucans, ergothioneine, flavonoids, ergosterols, zaragozic acids and phenolic compounds, which function as nutraceuticals via numerous mechanisms to lower CVD risk. This occurs, in part, by lowering total plasma and low-density lipoprotein (LDL) cholesterol levels, increasing bile acid excretion and hepatic LDL receptor mRNA expression and inhibiting HMG CoA reductase and squalene synthesis. Mushrooms also exert anti-inflammatory, immunomodulating, anti-coagulant, anti-hypertensive and anti-adhesive properties that collectively diminish atherosclerotic risk. Furthermore, dietary mushrooms may inhibit development of atherosclerotic plaques and purportedly reduce the overall plaque size if already present. Emerging research suggests

[*] Corresponding author: Keith R. Martin, PhD, MTox, Nutrition Program, Healthy Lifestyles Research Center, School of Nutrition and Health Promotion, Arizona State University, 6950 East Williams Field Road, Mesa, AZ 85212, Telephone: (480) 727-1925, Fax: (480) 727-1064, E-mail: Keith.R.Martin@asu.edu.

that many of these effects not only occur at the level of molecular interaction and enzymatic inhibition but also at the level of gene expression changes. Thus, dietary mushrooms are a rich source of nutraceuticals and may be an important dietary means of reducing the risk for CVD.

Keywords: cardiovascular disease, dietary mushrooms, nutrigenomics, nutraceuticals, bioactive agents

Introduction

Cardiovascular disease (CVD) is the leading cause of morbidity and mortality in the United States as well as globally [1] and nearly 1 in every 3 deaths each year can be attributed to CVD [2]. CVD encompasses any disease of the cardiovascular system that is related to atherosclerosis, which often is asymptomatic until it causes myocardial infarction, sudden cardiac death or stroke. Atherosclerosis is a condition in which arterial walls thicken and harden over time due to deposits of cholesterol within macrophage foam cells in the vascular intima. Chronic inflammation occurs and is exacerbated by excessive low density lipoprotein (LDL) cholesterol that penetrates the endothelial monolayer of the artery and becomes minimally modified or fully oxidized (Figure 1), increases production of pro-inflammatory, chemotactic cytokines and chemokines, and induces the expression of adhesion molecules that signal monocytes and leukocytes to bind to the site of inflammation [3].

Monocytes, a type of white blood cell, bind to adhesion molecules such as vascular cell adhesion molecule (VCAM-1), and migrate from the blood through the endothelial monolayer of the artery to the arterial intima where they differentiate into macrophages and actively engulf minimally modified and oxidized LDL producing cholesterol-laden foam cells.

These foam cells are then covered by fibrin and elastin to form fatty lesions called plaques, which reduce arterial elasticity and lumen cross-sectional area thereby restricting blood flow. Subsequently, plaque rupture can produce thrombi, which is the final common pathological pathway that leads to myocardial infarction or stroke.

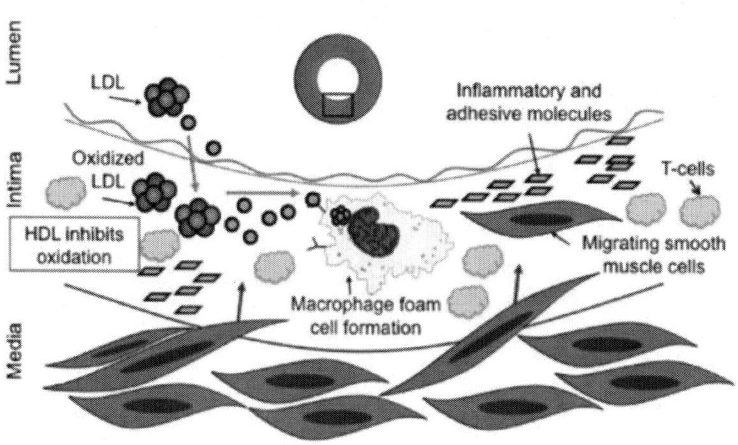

Gotto, J. Family Practice, 58(10): 58. Reprinted with permission.

Figure 1. Progression of atherosclerosis.

Plasma cholesterol levels and chronic inflammation are directly correlated and play significant roles in the risk and development of atherosclerosis [3]. Various dietary mushroom species have been linked to decreased plasma cholesterol and triglyceride levels, ameliorated hypertension and a reduction in size and incidence of aortic atherosclerosis in animal models [4-8].

The purpose of this review is to familiarize the reader with the beneficial effects of consuming edible mushrooms on cardiovascular disease risk with particular interest in the specific bioactive compounds in different species and the mechanisms of action associated with the beneficial effects.

Dietary Mushrooms and Nutraceuticals

For many years, mushrooms have been used to treat a variety of common ailments in traditional therapies. Mushrooms, by definition, are a macrofungi with distinctive and visible fruiting bodies that may grow above or below ground [9]. Recent estimates suggest that there are approximately 150,000 species of mushroom on earth, of which only 10% have been identified and named [10]. Many of these species are included as an important ingredient in a variety of culinary dishes due to their flavorful taste and nutritional value. In terms of nutrition, mushrooms are high in protein, fiber and a variety of essential vitamins and minerals each obviously with biological activity [11].

In addition, mushrooms contain a plethora of non-nutritional bioactive compounds that exert clear beneficial effects on cardiovascular health [12-14]. For this reason, many of these bioactive compounds have been formulated as purified single agents, extracts, or complex mixtures and marketed as "nutraceuticals" in many dietary supplements. Nutraceuticals, a hybrid term derived from nutrition and pharmaceutical, are foods or isolated food ingredients that aid in the prevention or treatment of disease [15].

In recognition of the numerous biological activities of mushrooms, the dietary supplement industry has coined the phrase "mushroom nutraceuticals" referring typically to a refined capsule-enclosed dry powder with intended therapeutic, medicinal applications [14]. These applications are generally thought to involve effects on the production and activity of biological proteins in key metabolic pathways due, in large part, to differential changes in gene expression, which is dependent on the genetic makeup of individuals. However, it is important to note that these effects may be fairly heterogeneous across populations due to gene-diet interactions [16].

Environmental risk factors for CVD consist mainly of diet and exercise but other factors such as stress, smoking, air pollution and heavy metal toxicity can also play a role in relative risk. Some diets have been shown to significantly reduce CVD risk such as the Mediterranean diet and the Dietary Approaches to Stop Hypertension (DASH) diet, which are low in fat and sodium and rich in cardioprotective fruits and vegetables, as well as monounsaturated and omega-3 polyunsaturated fats [17-19].

However, the benefit of these diets can vary markedly depending on an individual's DNA and subsequent diet-gene interactions [20]. Polymorphisms in genes that are involved in lipoprotein metabolism may ultimately determine the individual's risk of CVD, in terms of circulating cholesterol, LDL-C, HDL-C, triglycerides and lipoproteins [16].

Mushroom Bioactive Compounds and Mechanisms of Action

Mushrooms are replete with bioactive compounds that have been shown to exert hypocholesterolemic [5, 7, 21-25], immunomodulatory [26-28], anti-atherogenic [6, 7, 29-33], antioxidant [34-36], anti-hypertensive [7, 8, 37] and hypoglycemic [23] properties. Many of the mechanisms of action involve disruption of the atherosclerotic process generally by reducing atherogenic plasma lipids/lipoproteins, reducing lipid and bile acid reabsorption, inhibiting key, often rate-limiting, enzymes and inhibiting bile acid synthesis.

Hypocholesterolemic Effects

Dietary fiber encompasses the edible parts of plants or fungi that are resistant to digestion and absorption by humans. Increased consumption has been shown to markedly reduce plasma cholesterol levels. Moreover, accumulating evidence supports a lower incidence of CVD in persons consuming diets high in fiber. Dietary mushrooms are particularly rich in dietary fiber with beta-glucans, glucose polysaccharides linked by beta glycosidic bonds, representing 4-13% of total dietary fiber within mushrooms, depending on the species, and appearing primarily responsible for the observed hypocholesterolemic effects [13]. Two mushroom species, *Auriculae auricula* and *Tremella fuciformis*, are particularly high in dietary fiber [38], while other species such as *Agaricus bisporus* and *Pleurotus ostreatus* are moderately rich in dietary fiber. Soluble dietary fiber lowers total plasma and LDL cholesterol [39] through chemical interaction and formation of viscous gels (beta-glucans) that inhibit cholesterol and triglyceride absorption by binding to bile acids and hindering micelle formation [40]. Soluble fiber also decreases enterohepatic recirculation and increases fecal excretion of bile acids. The mechanism of action is similar to bile acid sequestrants such as cholestyramine or colestipol. The decrease in hepatic cholesterol promotes removal of LDL from the blood and the reduction in bile acids necessitates the use of cholesterol to synthesize more bile acids. In this way, plasma LDL and total cholesterol are beneficially lowered [41, 42].

In animal studies, changes in plasma lipids are accompanied by changes in gene expression. Rats fed with *Tremella fuciformis* dietary fiber for 4 weeks showed significant decreases in serum LDL, hepatic total cholesterol and triglycerides [43]. Total bile acid and short chain fatty acid (SFCA) excretion increased, suggesting a link to the therapeutic effects of soluble fiber present in mushrooms. Fukushima and coworkers (2000) also discovered significantly lower serum total cholesterol, as well as VLDL, intermediate-density lipoprotein (IDL) and LDL-C when they fed rats *Agaricus bisporus* fiber and compared against a control group [44].

The authors also noticed the same outcome on cholesterol levels when they fed rats maitake (*Grifola frondosa*) and enokitake (*Flammulina velutipes*) mushrooms [45]. The mechanism of action for both studies was linked to increased hepatic LDL receptor mRNA expression with presumably increased translation and synthesis of LDL receptor protein. The *Agaricus bisporus* and *Flammulina velutipes* groups both showed significantly higher LDL receptor mRNA expression when compared to a control group [44, 45]. These data suggest an effect by dietary mushrooms on gene expression.

Other fungal polysaccharide polymers with hypocholesterolemic properties similar to beta-glucan are chitin and chitosan [13]. Chitin is a long-chain polymer of an N-acetylglucosamine, a derivative of glucose, and is found in the cell walls of fungi. Chitosan is a linear polysaccharide composed of randomly distributed β-(1-4)-linked D-glucosamine (deacetylated unit) and N-acetyl-D-glucosamine (acetylated unit) formed from the deacetylation of chitin. Obese mice fed a high-fiber diet supplemented with a 5% fungal chitosan from *Agaricus bisporus* for 10 weeks displayed reduced lipid absorption and serum adipocytokine levels, when compared to obese mice fed a high fiber diet only [46]. The suppression of adipocytokines potentially suggests, in part, altered production of proteins, which could be due to alterations in respective gene expression.

Many species of dietary mushrooms reduce CVD risk by lowering plasma cholesterol thorough enzymatic inhibition. For example, the oyster mushroom (*Pleurotus ostreatus*) naturally produces the bioactive agent lovastatin [47-50], which has been shown to lower cholesterol [6, 7, 23] and reduce the number and size of atherosclerotic lesions [6, 7]. Lovastatin is an HMG CoA reductase inhibitor (the rate-limiting enzyme in *de novo* cholesterol biosynthesis), used as a hypolipidemic agent in those with hypercholesterolemia. Rodent studies revealed that *Pleurotus ostreatus* decreases VLDL, LDL and total cholesterol, as well as increasing the fractional catabolic rate of cholesterol. Rodent studies support this theory, revealing that *Pleurotus ostreatus* decreases the activity of HMG CoA reductase in rat liver cells.

Hossain and coworkers (2004) showed similar results to other animal studies, however, decreases in LDL, total cholesterol and triglycerides were limited to hypercholesterolemic rats only with no changes occurring in normocholesterolemic rats [31].

Interestingly, this effect has also been observed in hypercholesterolemic human subjects [23, 36] as well as diabetic patients who showed improvements in LDL, total cholesterol and triglycerides, with no toxic side effects to the liver or kidneys [23].

In the latter study, the effects were noted in both normo and hypercholesterolemic, diabetic patients. Kajaba and coworkers (2008) also demonstrated significant reductions in total plasma cholesterol and triglycerides in dyslipidemic patients [36]. Rabbits fed *Pleurotus florida* also showed improvement in HDL/total cholesterol and HDL/LDL cholesterol risk ratios, important CVD risk factors, primarily through reductions in LDL [51]. The data support a pan-species effect by *Pleurotus* (*ostreatus and florida*) on hypercholesterolemia making these mushrooms efficacious nutraceutical sources for reducing CVD risk.

The first to demonstrate the cholesterol-lowering effects of mushrooms was a group of researchers from Japan in 1966, who demonstrated that rats fed a dietary supplement of shiitake mushroom (*Lentinus edodes*) for 10 weeks, significantly lowered plasma cholesterol levels [21]. The bioactive agent presumably responsible for these hypocholesterolemic effects was identified as lentysine, which is now commonly referred to as eritadenine, an acyclic sugar adenosine analogue or polysaccharide beta-glucan [4, 5, 52-57]. Since the initial discoveries, the results have been corroborated many times indicating that total plasma cholesterol levels are decreased in rats fed eritadenine [53-57].

In subsequent studies, shiitake mushrooms (*Lentinus edodes*) were shown to have cholesterol-lowering effects in humans [22]. Although eritadenine is found mainly in shiitake mushrooms (*Lentinus edodes*) [21, 52, 58], it occurs in lesser amounts in white button mushrooms (*Agaricus bisporus*) [4].

A mechanism has been proposed for the hypocholesterolemic properties of eritadenine. It is theorized to be a decrease in the phosphatidylcholine (PC)/phosphatidylethanolamine (PE) ratio [52]. Both PC and PE are major phospholipids in mammalian plasma membranes and the ratio is important in steatosis.

Mechanistically, eritadenine is a potent inhibitor of the enzyme S-adenosyl-homocysteine hydrolase, causing an increase in s-adenosyl-homocysteine concentrations. This causes a change in the composition of membrane phospholipids, which causes an up-regulation of lipoprotein receptors in the liver and consequently an increased uptake of plasma cholesterol [57]. The cholesterol and LDL-lowering effects could explain the success of eritadenine-rich mushrooms as a frequently purchased and used nutraceutical.

Some mushroom species also produce cholesterol-inhibitors known as zaragozic acids, [24, 25] which are a family of naturally occurring fungal products possessing a unique 4, 8-dioxabicyclo (3.2.1) octane structure, with variation in the 1-alkyl and 6-acyl side chains. These compounds are potent inhibitors of squalene synthase, which is the first committed enzymatic step in sterol synthesis and ultimately cholesterol synthesis [24]. As such, consumption of zaragozic acid lowers plasma cholesterol levels. In mechanistic studies, rats fed zaragozic acid displayed an increase in hepatic low density lipoprotein (LDL) receptor mRNA levels coupled with marked increases in HMG CoA synthase, squalene synthase, and HMG CoA reductase, which appeared responsible for increases in enzyme protein and activity.

The authors suggest in this study that inhibition of squalene synthase decreased the level of a squalene-derived regulatory product, resulting in transcriptional modulation of several mRNAs and subsequent increases in HMG CoA reductase mRNA, protein, and activity. Moreover, the increase in HMG CoA reductase gene expression was correlated with the inhibition of cholesterol synthesis caused by zaragozic acid.

Zaragozic acids have also been shown to inhibit ras-farnesyl-protein transferase (FTI). This enzyme is important in the post-translational modification and activation of the ras protein, a G protein anchored in the plasma membrane and involved in second messenger signaling [25]. Ras may be involved in smooth muscle cell (SMC) proliferation or senescence noted in plaque development in atherosclerosis. Interestingly, statins not only have favorable effects on plasma lipids but also appear to prevent the proliferation of vascular SMC that is characteristic of atherosclerosis. This may be due to modulation of Ras, which has been shown to induce vascular SMC senescence, inflammation, and other age-related vascular problems [59]. Sugita et al. (2007) examined the effects of an FTI, viz., manumycin A, on the development and progression of atherosclerosis in ApoE-deficient mice fed a high-fat diet [60]. The authors found a significant, reduced development of mature atherosclerosis with concomitant, marked reduction in oxidative stress in mice consuming the FTI. The collective data suggest that Ras or proteins in the downstream signaling cascade may be valuable molecular targets for mushroom-derived nutraceuticals in reducing or ameliorating CVD risk.

Mushrooms rich in phytosterols such as ergosterol may lower plasma cholesterol by a mechanism other than disruption of micelles and reduced cholesterol absorption. Ergosterols are biological precursors to vitamin D2, being converted by ultraviolet light to ergocalciferol. Vitamin D deficiency is prevalent in the US and estimated to be present in ~30-50% of the general population. Moreover, vitamin D2 deficiency has been associated with hypertension, obesity, diabetes, and the metabolic syndrome as well as stroke and congestive heart failure even after adjustment for traditional CVD risk factors [61, 62].

It is also associated with markers of subclinical atherosclerosis including increased intima-media thickness and decreased coronary circulation [63, 64]. One proposed mechanism is that deficiency causes increases in parathyroid hormone, which increases insulin resistance associated with diabetes, hypertension, inflammation, left ventricular hypertrophy, and increased cardiovascular risk [65, 66]. As such, vitamin D deficiency has been identified as a novel CVD risk factor due to its striking correlation with myriad cardiovascular events. This renders dietary mushrooms a unique and novel nutraceutical source of vitamin D in decreasing CVD risk.

Ganoderma lucidum, also known as lingzhi or reishi, contain ganoderic acids, which are triterpenoids (derivatives from lanosterols) with molecular structures similar to steroid hormones that also have been shown to inhibit cholesterol synthesis [67]. Researchers have isolated 26-oxygenosterol from *Ganoderma lucidum*, a ganoderic acid that inhibits lanosterol 14α-demethylase-the enzyme that converts 24, 25-dihydrolanosterol to cholesterol, thus decreasing conversion of acetate or mevalonate as precursors to cholesterol synthesis and subsequently bile acid synthesis [68].

Given these effects, *Ganoderma* represents a key nutraceutical source of compounds that inhibit biosynthesis of cholesterol. Mushroom consumption has been shown to have fairly potent hypolipidemic effects in humans as well [23, 69]. Agrawal et al. (2010) demonstrated that oyster mushroom ingestion over a 3-month period lowered total cholesterol levels by 16%, VLDL by 26%, LDL-C by 6% and triglycerides by 25% while raising HDL-C levels by 8% in patients with type 2 diabetes.

Similarly, Khatun and colleagues (2007) demonstrated significant reductions in serum triglycerides and cholesterol after 7-days of mushroom consumption in type 2 diabetic patients. The various cholesterol-lowering effects of mushrooms are summarized in Figure 2 (below), which shows the physiological pathways that are altered by bioactive compounds.

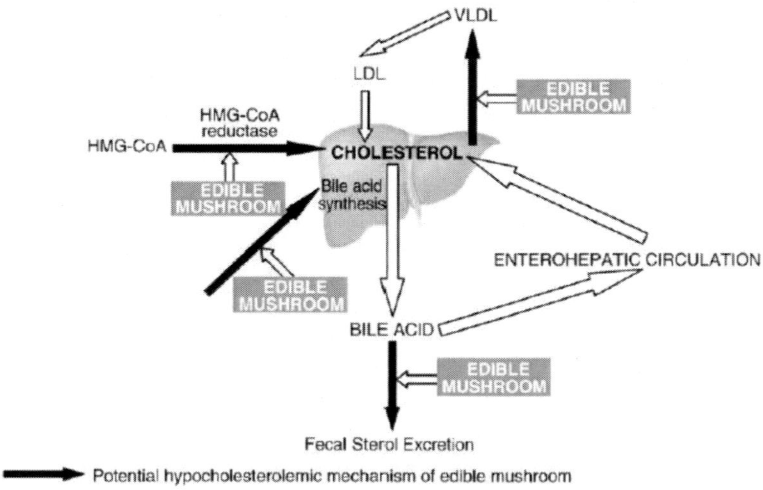

Reprinted with permission from E. Guiillamon et al./Fitoterapia 81 (2010) 715-723.

Figure 2. Effects of edible mushrooms on cholesterol metabolism.

Inhibition of Vascular Adhesion Molecule Expression

The etiology of CVD involves up-regulation of endothelial adhesion molecules, binding of monocytes, extravasation into the aortic intima, and foam cell formation associated, in part, with inflammation [70, 71].

Pro-inflammatory cytokines such as IL-1β induce several key adhesion molecules (intercellular adhesion molecule-1 (ICAM-1), vascular cell adhesion molecule-1 (VCAM-1), and endothelial-leukocyte adhesion molecule-1 (ELAM-1 or E-selectin) and contribute significantly to atherosclerotic plaque development, premature cardiovascular disease and death.

In several studies, mushroom extracts and their purified agents inhibited adhesion molecule expression and binding of monocytes to the aortic endothelium [72, 73]. In one study, several mushroom species and types (*Lentinus edodes, Grifola frondosa, Pleurotus ostreatus, Agaricus bisporus*, [white button, crimini and portabella mushrooms]) significantly decreased adhesion molecule expression of VCAM-1, ICAM-1 and E-selectin, and subsequently reduced binding of monocytes in an in vitro model using non-transformed human aortic endothelial cells that had been placed in a pro-inflammatory environment [72]. In another study using ergothioneine, a purified bioactive agent from mushrooms, adhesion molecule expression was again significantly reduced under identical experimental conditions [73]. Moreover, ergothioneine significantly reduced the binding of human monocytes to the aortic endothelium after stimulation with pro-inflammatory cytokine interleukin 1β. Others have shown in an animal model that *Lentinus edodes* decreased the surface involvement of adhesion molecules and reduced the number of foam cells in the intima of aortic endothelium in rabbits possibly via the effects of the eritadenine molecule [30]. Xu et al. investigated the effect of polysaccharides from *Lentinus edodes* on the expression of VCAM-1 mRNA in endothelial cells of thoracic aorta endothelial cells in pro-atherogenic, high-fat-diet rats. *Lentinus edodes* significantly reduced atherogenic plasma lipids. In rats fed *Lentinus edodes* and provided the high fat diet, VCAM-1 mRNA levels were significantly reduced compared to controls [74]. This indicates that mushrooms were reducing gene expression of a key adhesion molecule involved, in part, with the formation of atherosclerotic plaques in aortic endothelium.

Regression of Atherosclerotic Plaques

Mushrooms have also been shown to inhibit plaque formation and cause regression in pre-existing lesions [6, 7]. A 10% oyster mushroom (*Pleurotus ostreatus*) diet significantly reduced both the incidence and size of aortic atherosclerotic plaques in rabbits suggesting interaction with physiological processes of atherogenesis [6, 7]. *Pleurotus ostreatus* significantly reduced the incidence of atherosclerotic plaques in rabbits and plaques by 26% compared to the control group and prevented the formation of atheroma, reduced foam cell numbers, and reduced coronary arterial injury [7]. Percario and coworkers (2008) also showed that a natural antioxidant-rich mushroom, *Agaricus sylvaticus*, can prevent the development of atherosclerosis in a hypercholesterolemic environment using rabbit models

[75]. The Atherosclerotic Index (AI), a measure of plaque size and severity, also indicated improvements after supplementation of *Lentinus edodes* in the same study. *Pleurotus eryngii, Grifola frondosa* and *Hypsizygus marmoreus* fed to atherosclerotic-susceptible, apolipoprotein E deficient mice caused a significant decrease in atherosclerotic lesions when compared to control [33].

Anti-Inflammatory and Vascular Effects

The association between chronic low-grade systemic inflammation and incident CVD is well-established [76-78]. Chronic low-grade inflammation is characterized by a 2-4 fold increase in systemic concentrations of cytokines such as tumor necrosis factor-α (TNF-α) and interleukin – 6 (IL-6) [79]. This inflammatory state is associated with a deleterious vascular milieu that is characterized by endothelial dysfunction, abnormal glucose metabolism and insulin resistance which are in turn associated with progression of atherosclerotic disease.

TNF-α is a significant player in the pathogenesis of insulin resistance and metabolic syndrome [80, 81]. It has been shown to impair insulin-stimulated glucose uptake in cell culture and animal studies. In humans, TNF-α infusion has been shown to adversely affect the insulin signaling cascade that regulates glucose transporter type-4 (GLUT-4) translocation and glucose uptake [82]. Thus, TNF-α may represent the molecular link between chronic inflammation, insulin resistance and atherosclerotic disease.

Preliminary data suggest that oyster mushrooms (*Pleurotus ostreatus*) may have significant anti-inflammatory activity by inhibiting TNF-α secretion [83]. Jedinak and colleagues examined the anti-inflammatory effects of oyster mushrooms in vitro and in vivo following lipopolysaccharide (LPS) infusion. LPS is a component of the gram-negative bacterial cell wall, and is known to be a potent inducer of TNF-α production. The authors found that oral administration of oyster mushroom extract suppressed secretion of TNF-α and IL-6 in mice challenged with LPS. The oyster mushroom concentrate also appeared to inhibit LPS-induced production of prostaglandin E2 and nitric oxide through cyclooxygenase-2 and inducible nitric oxide synthase down-regulation, respectively.

Mushrooms have other significant anti-inflammatory properties as well [26, 29]. A methanol extract of *Pleurotus florida* significantly inhibited acute inflammation induced by carrageenan and chronic inflammation induced by formalin in mice [29]. The *Pleurotus florida* extract also exhibited significant inhibitory effects on adenosine 5'-diphosphate (ADP)-induced platelet aggregation in washed human platelets.

In addition, a variety of mushroom species (*Lentinus edodes, Grifola frondosa, Pleurotus ostreatus, Agaricus bisporus*) decreased the binding of monocytes when in-vitro, human aortic endothelial cells that had been placed in a pro-inflammatory environment [72]. Furthermore, white button mushrooms (*Agaricus bisporus*) decreased the expression of VCAM-1, ICAM-1 and E-selectin. A significant reduction of adhesion molecule expression and the ultimate binding of these adhesion molecules to the endothelium show potential to decrease CVD risk. Thus, preliminary data from cell culture and animal studies suggest that mushrooms may have significant anti-inflammatory effects that could affect atherosclerotic disease progression.

The decreases in plaque formation with mushroom supplementation are promising as well. However, all the above findings need to be confirmed in well-controlled human studies.

Antioxidant Activity

The oxidation or modification of LDL plays an important role in the formation of atherosclerotic plaques in arteries since this altered form is freely accumulated by macrophages in the aortic intima resulting in foam cell formation [3]. Mushrooms possess a variety of antioxidant compounds such as polysaccharides (beta-glucans), ergosterol, nicotinic acid, triterpenes and phenolic compounds, which can interrupt this pro-oxidative process [11, 13, 52]. In addition, some mushroom species are also rich in vitamin A (beta-carotene), vitamin C and vitamin E (alpha-tocopherol). Collectively, dietary mushrooms are a reasonably good source of antioxidants that could reduce the risk of CVD.

The mushroom species *Agrocybe aegerita* has shown strong free radical-scavenging activity and in vitro studies have revealed an inhibitory effect on human LDL oxidation [84]. Lo and coworkers (2005) also noticed a positive correlation between total phenolic content and antioxidant activity, suggesting that the phenolic compounds present in mushrooms contribute to antioxidant potential. *Agrocybe aegerita* also shows cyclooxygenase inhibitory activity, which decreases inflammation through the decreased synthesis of pro-inflammatory prostaglandins [35]. The mechanism of action is similar to anti-inflammatory drugs such as aspirin or ibuprofen. Kajaba and coworkers (2008) found that *Pleurotus ostreatus* increases the activity of glutathione peroxidases (family of antioxidant enzymes) in dyslipidemic humans [36], which inactivates lipid hydroperoxides and hydrogen peroxide. These results were also corroborated by a decrease in conjugated dienes, indicators of enhanced lipid peroxidation. Researchers have also shown that supplementing *Lentinus edodes* to rats on a high-fat diet increased serum antioxidant enzyme activity and presumably antioxidant status [74]. Mushrooms also contain a number of enzymes such as lactase, superoxide dismutase, glucose oxidase and peroxidase, which help to reduce oxidative damage [14].

Ergothioneine is a very stable and potent antioxidant that is produced in nature only by fungi and a few Mycobacteria. Mushrooms are a superior dietary source of the same. Ergothioneine is an amino acid derivative of histidine and is not synthesized in humans, but levels increase in human blood as a result of dietary consumption. In vitro studies demonstrate that ergothioneine inactivates numerous, diverse free radicals and reactive oxygen species making it potentially one of the most effective hydrophilic bioactive molecules [73]. Interestingly, ergothioniene is the only dietary antioxidant that has its own genetically coded transporter in humans suggesting a unique antioxidative function. Recent research indicates that it protects developing erythrocytes against damage from ferryl hemoglobin, a highly reactive intermediate in auto-catalytic oxidation that leads to heme degradation [85].

Selenium is required for health and is considered an indirect antioxidant based on its incorporation into glutathione peroxidases, a family of enzymes that inactivate pro-oxidative lipid hydroperoxides and hydrogen peroxide [86]. Oxidative stress has been shown to be a significant source of blood vessel damage in CVD and has been correlated with low intake of selenium. As a result, efforts have been initiated to use white button mushrooms as a nutraceutical source for organic selenium (as selenomethionine) by supplementing commercial compost with selenium selenite (1200 ppm) at the spawning developmental phase. It has been reported that selenium concentrations of 30-300 ppm can be produced approaching 100% of the RDA for selenium in a single serving of mushrooms. Consequently, efforts are being made to develop selenium-rich mushrooms as a dietary supplement or as a value-added ingredient for the formulation of other functional foods and nutraceuticals [87].

Glucose Metabolism and Insulin Action

Type 2 diabetes and insulin resistance are strong predictors of cardiovascular disease risk. Insulin resistant and diabetic individuals have an elevated risk of adverse cardiovascular events [88-90]. Some mushroom species have been shown to have hypoglycemic, insulin-like or insulin releasing activity [91, 92, 69, 23]. Gray et al. (1998) examined the effects of *Agaricus campestris* administration on streptozotocin-diabetic mice [91]. Administration of the aqueous extract resulted in a 50% improvement in glucose oxidation, and 80% improvement in glycogen storage in mouse abdominal muscle. Glucose uptake improved by 96% as well. Baseline blood glucose levels in the mice treated with mushroom extract were normalized by day 12 of the study. Addition of the extract to the BRIN-BD11 pancreatic ß cell line resulted in a 3.5-4.6 fold stimulation of insulin secretion.

Similarly, Manohar et al. (2002) examined the effects of an aqueous extract of maitake mushroom (Fraction X (FXM)) on circulating glucose and insulin concentration in KK mice [92]. Insulin resistant KK mice were gavaged with a single dose of varying concentrations of FXM, glipizide (an insulin sensitizing agent) and placebo. FXM resulted in a 25% reduction in circulating blood glucose levels and was comparable to reductions achieved by glipizide administration (25-37%). Circulating levels of insulin following chronic administration of FXM were significantly lower as compared to the glipizide group that showed no significant changes.

Preuss and co-workers (2010) examined the effects of a specific glycoprotein extract of maitake mushrooms (SX fraction) on insulin sensitivity in spontaneously hypertensive rats [37]. The effect of the SX fraction on insulin sensitivity was compared to the insulin sensitizing agent pioglitazone. The authors reported that consumption of sucrose with medium and high-levels of maitake extract was comparable in efficacy to pioglitazone administration.

Agrawal and colleagues (2010) conducted a randomized controlled trial to examine the effects of oyster mushroom consumption on glycemia and long term-glycemic control in patients with type 2 diabetes [69].

At the end of the 3-month long intervention, fasting blood glucose and HbA1c were reduced by 47% and 13%, respectively. It is important to note, that these salutary changes occurred in the absence of significant weight loss. Similarly, Khatun and colleagues (2007) observed significant reductions in fasting plasma glucose following oyster mushroom consumption in type 2 diabetic subjects [23].

Many putative mechanisms have been proposed to explain the anti-glycemic activity of mushrooms. Beta-glucans found in mushrooms are thought to lower blood sugar levels by slowing the absorption of sugar [92]. Guanidine which belongs to the biguanide class of oral hypoglycemic agents has also been detected in edible mushrooms [91] and may play a role in explaining the observed glucose lowering effects.

At present however, the specific compound within mushrooms that causes anti-glycemic effects remains to be found and isolated.

Improvements in Blood Pressure

High blood pressure (>140/90 mmHg) is a major contributor to cardiovascular disease risk. Data from observational studies involving >1 million patients have indicated that CVD risk rises monotonically at blood pressures >115/75 mmHg with a doubling of risk for every increment of 20/10 mmHg [93]. As per current Joint National Committee guidelines, lifestyle modification is an important adjunct to pharmacotherapy for blood pressure management.

Several studies have also shown that mushrooms have hypotensive properties [5, 8, 23, 37]. Maitake mushrooms (*Grifola frondosa*) in particular, seem to ameliorate hypertension in rats [5, 8, 37]. Talpur and coworkers (2003) observed decreases in systolic blood pressure when they fed aged, diabetic Zucker rats maitake mushroom fractions [8]. Shiitake mushrooms (*Lentinus edodes*) have also been shown to reduce blood pressure in hypertensive rats [5]. Preuss and coworkers (2010) also found that maitake mushrooms stopped the age-related increase in blood pressure in aging, female rats [37].

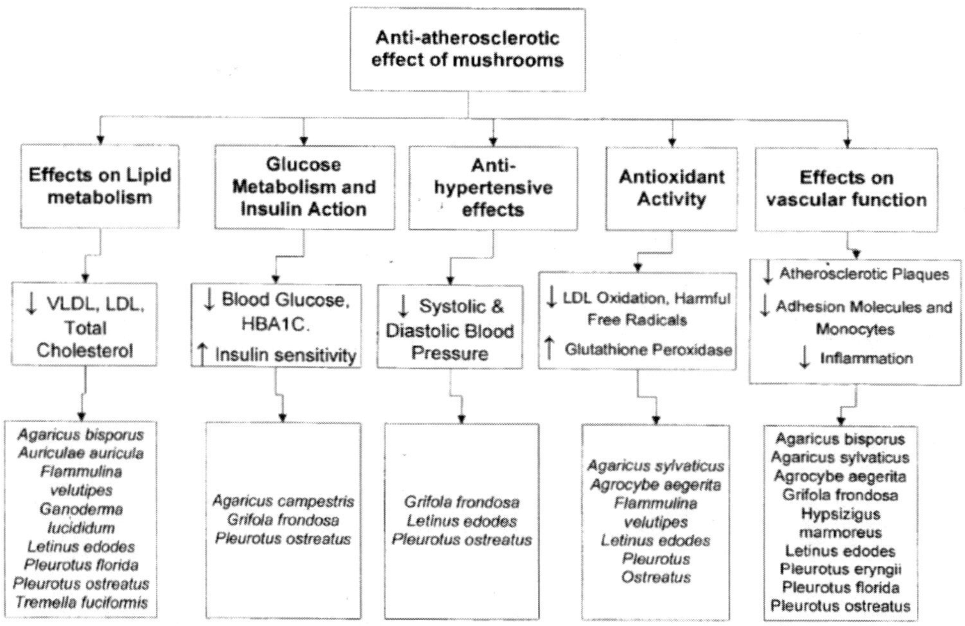

Figure 3. Anti-atherosclerotic effect of mushrooms.

Agrawal and colleagues (2010) examined the effects of oyster mushroom consumption on blood pressure in patients with type 2 diabetes [69]. They observed statistically significant reductions of 5 mmHg and 2 mmHg in systolic and diastolic blood pressure respectively at the end of the 3-month intervention period. Similarly, Khatun and coworkers (2007) found that feeding oyster mushrooms (*Pleurotus ostreatus*) significantly decreased the blood pressure of middle-aged, type 2 diabetics [23].

Mushrooms may lower blood pressure by decreasing the activity of serum ACE and the renin-angiotensin system (RAS) [37]. It is also plausible, that the insulin sensitizing effects of mushroom consumption may play a role in blood pressure reduction as well.

In humans, the observed reductions in systolic (-5 mmHg) and diastolic (-2 mmHg) pressures may have significant implications for cardiovascular risk reduction since a reduction in systolic blood pressure of 5 mmHg has been associated with a 14% reduction in the incidence of stroke, 9% reduction in the incidence of coronary heart disease and a 7% reduction in total mortality [94].

Applications in Modern Medicine and Dietary Supplements

Clearly there are numerous bioactive agents in fungi that can reduce the risk for CVD. In fact, many of these compounds are currently the drugs used for CVD risk including lovastatin and bile acid sequestrants. There are numerous other bioactive agents that have shown remarkable promise including eritadenine and zaragozic acid. Collectively, this makes dietary mushrooms a robust and attractive source of nutraceutical compounds for reducing the risk of CVD.

Conclusion

Mushrooms have cardio-protective properties linked to a host of bioactive compounds that they possess. Eritadenine, lovastatin, beta-glucans and zaragozic acids lower total plasma and LDL cholesterol levels via a variety of physiological mechanisms. Increases in bile acids, inhibition of HMG CoA reductase and increases in expression of hepatic LDL receptor mRNA cause these hypocholesterolemic effects. Mushrooms have also been shown to reduce inflammatory processes associated with athlerosclerotic plaque formation. Oyster mushrooms (*Pleurotus ostreatus*) in particular have been shown to decrease the size and incidence of plaque formation in rabbit and rodent aorta. The high antioxidant activity of mushrooms also helps to decrease the level of oxidized-LDL in the plasma, which decreases the formation of athlerosclerotic plaques typically induced by inflammatory processes such as monocyte binding and formation of foam cells. Finally, several species of mushrooms also inhibit the expression of aortic adhesion molecules, which decrease the binding of monocytes and consequently the deposition of plaques in the arteries. Based on available research, mushrooms seem to have therapeutic effects for individuals that are at an increased risk of CVD.

References

[1] CDC 2007 Mortality Data. http://www.cdc.gov/nchs/fastats/deaths.htm. Accessed 17 July 2011.

[2] Roger, V. L., Go, A. S., Lloyd-Jones, D. M., Adams, R. J., et al. Heart disease and stroke statistics—2011 update: a report from the American Heart Association. *Circulation*. 2011; 123(4): e18-e209. Epub. 2010 Dec. 15.

[3] Libby, P. Vascular biology of atherosclerosis: Overview of state of the art. *Am. J. Cardiol*. 2003; 91 (suppl.): 3A-6A.

[4] Mattila, P., Suonpaa, K., Piironen, V. Functional properties of edible mushrooms. *Nutrition*. 2000; 16(7):694-696.

[5] Kabir, Y., Yamaguchi, M., Kimura, S. Effect of shiitake (Lentinus edodes) and maitake (Grifola frondosa) mushrooms on blood pressure and plasma lipids of spontaneously hypertensive rats. *J. Nutr. Sci. Vitaminol*. 1987; 33(5): 341-346.

[6] Bobek, P., Galbavy, S. The oyster mushroom (Pleurotus ostreatus) effectively prevents the development of atherosclerosis in rabbits. 48. 1999; 5:226-230.

[7] Bobek, P., Galbavy, S. Hypocholestolemic and antiatherogenic effect of oyster mushroom (Pleurotus ostreatus) in rabbits. *Nahrung*. 1999; 43:339-342.

[8] Talpur, N., Echard, B., Yasmin, T., et al. Effects of niacin-bound chromium, Maitake mushroom fraction SX and (-)-hydroxycitric acid on the metabolic syndrome in aged diabetic Zucker fatty rats. *Mol. Cell. Biochem*. 2003; 253(1-2):369-377.

[9] Miles, P. G., Chang, S. T. *Mushroom Biology: Concise Basics and Current Developments*. Singapore: World Scientific; 1997.

[10] Kirk, P. M., Cannon, P. F., David, J. C., et al. *Ainsworth and Brisby's dictionary of the fungi*, 10th Ed. CAB International, Wallingford. 2008.

[11] Mattila, P., Konko, K., Eurola, M., et al. Contents of vitamins, mineral elements, and some phenolic compounds in cultivated mushrooms. *J. Agric. Food Chem*. 2001; 49(5):2343-2348.

[12] Francia, C., Rapior, S., Courtecuisse, R., et al. Current research findings on the effects of selected mushrooms on cardiovascular diseases. *Int. J. Med. Mushrooms*. 1999; 1:169-172.

[13] Guillamon, E., Garcia-Lafuente, A., Lozano, M., et al. Edible mushrooms: Role in the prevention of cardiovascular diseases. *Fitoterapia*. 2010; 81:715-723.

[14] Wasser, S. P. Current findings, future trends, and unsolved problems in studies of medicinal mushrooms. *Appl. Microbiol. Biotechnol*. 2011; 89:1323-1332.

[15] Kalra, E. K. Nutraceutical-definition and introduction. *AAPS Pharm. Sci*. 2003; 5(3): E25.

[16] Engler, M. B. Nutrigenomics in cardiovascular disease implications for the future. *Prog. Cardiovasc. Nurs*. 2009; 24(4):190-195.

[17] Bemelmans, W. J., Broer, J., Feskens, E. J., et al. Effect of an increased intake of alpha-linoleic acid and group nutritional education on cardiovascular risk factors: the Mediterranean Alpha-linoleic Enriched Groningen Dietary Intervention (MARGARIN) study. *Am. J. Clin. Nutr*. 2002; 75: 221-227.

[18] Fuentes, F., Lopez-Miranda, J., Sanchez, E., et al. Mediterranean and low-fat diets improve endothelial function in hypercholesterolemic men. *Ann. Intern. Med.* 2001; 134: 1115-1119.

[19] Sacks, F. M., Svetkey, L. P., Vollmer, W. M., et al. Effects on blood pressure of reduced dietary sodium and Dietary Approaches to Stop Hypertension (DASH) diet. *N Engl. J. Med.* 2001; 344(1):3-10.

[20] Ordovas, J. M. Genetic interactions with diet influence the risk of cardiovascular disease. *Am. J. Clin. Nutr.* 2006; 83 (suppl.): 443S-446S.

[21] Kaneda, T., Tokuda, S. Effect of various mushroom preparations on cholesterol levels in rat. *J. Nutr.* 1966; 90:371.

[22] Suzuki, S., Ohshima, S. Influence of shiitake (Lentinus edodes) on human serum cholesterol. *Mushroom Sci.* 1976; 9:463.

[23] Khatun, K., Mahtab, H., Khanan, P. A., et al. Oyster mushroom reduced blood glucose and cholesterol in diabetic subjects. *Mymensingh. Med. J.* 2007; 16(1):94-99.

[24] Bergstrom, J., Dufresne, D., Bills, G., et al. Discovery, biosynthesis, and mechanism of action of the zaragozic acid: Potent inhibitors of squalene synthase. *Ann. Rev. Microbiol.* 1995; 49: 607-639.

[25] Dufresne, D., Wilson, K., Singh, S., et al. Zaragozic acids D and D2: Potent inhibitors of squalene synthase and Ras farnesyl-protein transferase. *J. Nat. Prod.* 1993; 56(11):1923-1929.

[26] Lull, C., Wichers, H. J., Savelkoul, H. F. Antiinflammatory and immunomodulating properties of fungal metabolites. *Mediators Inflamm.* 2005; 2005(2):63-80.

[27] Liu, Y., Fukuwatari, Y., Okumura, K., et al. Immunomodulating activity of Agaricus brasilienis KA21 in mice and in human volunteers. *Evidence-based Compl. Alt. Med.* 2008; 5(2):205-219.

[28] Kohno, K., Miyake, M., Sono, O., et al. Anti-inflammatory and immunomodulatory properties of 2-amino-3H-phenoxazin-3-one. *Biol. Pharm. Bull.* 2008; 31(10):1938-1945.

[29] Jose, N., Ajith, T. A., Janardhanan, K. K. Methanol extract of the oyster mushroom, Pleurotus florida, inhibits inflammation and platelet aggregation. *Phytother. Res.* 2004; 18(1):43-46.

[30] Yamada, T., Oinumo, T., Niihashi, M., et al. Effects of Lentinus edodes mycelia on dietary-induced atherosclerotic involvement in rabbit aorta. *J. Atheroscler. Thromb.* 2002; 9(3):149-156.

[31] Hossain, S., Hashimoto, M., Choudhury, E. K., et al. Dietary mushroom (Pleurotus ostreatus) ameliorates atherogenic lipid in hypercholesterolaemic rats. *Clin. Exp. Pharmacol. Physiol.* 2003; 30(7):470-475.

[32] Votruba, I., Holy,´A. Eritadenine, a novel type of potent inhibitors of S-adenosyl-L-homocysteine hydrolase. *Collect. Czech. Chem. Commun.* 1982; 47:167-172.

[33] Mori,K,. Kobayashi, C., Tomita, T., et al. Antiatherosclerotic effect of the edible mushrooms Pleurotus enryngii (Eringi), Grifola frondosa (Maitake), and Hypsizygus mormoreus (Bunashimeji) in apolipoprotein E-deficient mice. *Nutr. Res.* 2008; 28(5): 335-342.

[34] Hu, S. H., Liang, Z. C., Chia, Y. C., et al. Antihyperlipidemic and antioxidant effects of extracts from Pleurotus citrinopileatus. *J. Agric. Food Chem.* 2006; 54(6):2103-2110.

[35] Zhang, Y., Mills, G. L., Nair, M. G. Cyclooxygenase inhibitory and antioxidant compounds from the fruiting body of an edible mushroom, Agrocybe aegerita. *Phytomedicine*. 2003; 10(5):386-390.

[36] Kajaba, I., Simoncic, R., Frecerova, K., et al. Clinical studies on the hyperlipidemic and antioxidant effects of selected natural substances. *Bratisl. Lek. Listy*. 2008; 109(6):267-272.

[37] Preuss, H. G., Echard, B., Bagchi, D., et al. Maitake mushroom extracts ameliorate progressive hypertension and other chronic metabolic perturbations in aging female rats. *Int. J. Med. Sci*. 2010; 7:169-180.

[38] Cheung, P. C. K. The hypocholesterolemic effect of extracellular polysaccharide from the submerged fermentation of mushroom. *Nutr. Res.* 1996; 11:1953-1957.

[39] Erkkilla, A. T., Lichtenstein, A. H. Fiber and cardiovascular disease risk: how strong is the evidence. *J. Cardiovasc. Nurs*. 2006; 21(1):3-8.

[40] Gropper, S. S., Smith, J. L., Groff, J. L. *Advanced Nutrition and Human Metabolism*: 5th Edition. Wadsworth: Belmont. 2009.

[41] Alban, S., Franz, G. Partial synthetic glucan sulfates as potential new antithrombotics: a review. *Biomacromolecules*. 2001; 2(2):354-361.

[42] Zekovic, D. B., Kwiatkowski, S., Vrvic, M. M., et al. Natural and modified (1→3)-beta-D-glucans in health promotion and disease alleviation. *Crit. Rev. Biotechnol.* 2005; 25(4):205-230.

[43] Cheng, H. H., Hou, W. C., Lu, M. L. Interactions of lipid metabolism and intestinal physiology with Tremella fuciformis berk edible mushroom in rats fed a high-cholesterol diet with or without nebacitin. *J. Agric. Food Chem.* 2002; 50:7438-7443.

[44] Fukushima, M., Nakano, M., Morii, Y., et al. Hepatic LDL receptor mRNA in rats is increased by dietary mushroom (Agaricus bisporus) fiber and sugar beet fiber. *J. Nutr.* 2000; 130(9):2151-2156.

[45] Fukushima, M., Ohashi, T., Fujiwara, Y., et al. Cholesterol-lowering effects of Maitake (Grifola frondosa) fiber, Shiitake (letinus edodes) fiber, and Enokitake (Flammulina velutipes) fiber in rats. *Exp. Biol. Med.* 2001; 226(8):758-765.

[46] Neyrinck, A. M., Bindels, L. B., De Backer, F., et al. Dietary supplementation with chitosan derived from mushrooms changes adipocytokine profile in diet-induced obese mice, a phenomenon linked to its lipid-lowering action. *Int. Immunopharmacol.* 2009; 9(6):767-773.

[47] Gunde-Cimerman, N., Cimerman, A. Pleurotus fruiting bodies contain the inhibitor of 3-hydroxy-3-methylglutaryl-coenzyme A reductase-Lovastatin. *Exper. Mycol.* 1995; 19(1):1-6.

[48] Alarcon, J., Aguila, S., Arancibia-Avila, P., et al. Production and purification of statins from Pleurotus ostreatus (Basidiomycetes) strains. *Zeitschrift fur Naturforschung*. 2003; 58(1-2): 62-64.

[49] Bobek, P., Kuniak, L., Ozdin, L. The mushroom Pleurotus ostreatus reduces secretion and accelerates the fractional turnover rate of very-low-density lipoproteins in the rat. *Ann. Nutr. Metab.* 1993; 37(3):142-145.

[50] Bobek, P., Ozdin, L., Kuniak, L. Mechanism of hypocholesterolemic effect of oyster mushroom (Pleurotus ostreatus) in rats: reduction of cholesterol absorption and increase of plasma cholesterol removal. *Z Ernahrungswiss.* 1994; 33(1):44-50.

[51] Bajaj, M., Vadhera, S., Brar, A. D., et al. Role of oyster mushroom (Pleurotus florida) as hypocholesterolemic/antiatherogenic agent. *Indian J. Exp. Biol.* 1997; 35(10): 1070-1075.

[52] Enman, J., Rova, U., Berglund, K. Quantification of the bioactive compound eritadenine in selected strains of shiitake mushroom (Lentinus edodes). *J. Agric. Food Chem.* 2007; 55(4):1177-1180.

[53] Shimada, Y., Morita, T., Sugiyama, K. Eritadenine-induced alterations of plasma lipoprotein lipid concentrations and phosphatidylcholine molecular species profile in rats fed cholesterol-free and cholesterol-enriched diets. *Biosci. Biotechnol. Biochem.* 2003; 67:996-1006.

[54] Shimada, Y., Morita, T., Sugiyama, K. Dietary eritadenine and ethanolamine depress fatty acid desaturase activities by increasing liver microsomal phosphatidylethanolamine in rats. *J. Nutr.* 2003; 133:758-765.

[55] Shimada, Y., Yamakawa, A., Morita, T., et al. Effects of dietary eritadenine on the liver microsomal delta 6-desaturase activity and its mRNA in rats. *Biosci. Biotechnol. Biochem.* 2003; 67:1258-1266.

[56] Sugiyama, K., Akachi, T., Yamakawa, A. Eritadenine-induced alteration of hepatic phospholipid metabolism in relation to its hypocholesterolemic action in rats. *J. Nutr. Biochem.* 1995; 6:80-87.

[57] Sugiyama, K., Yamakawa, A., Kawagishi, H., et al. Dietary eritadenine modifies plasma phosphatidylcholine molecular species profile in rats fed different types of fat. *J. Nutr.* 1997; 127:593-599.

[58] Rokujo, T., Kikuchi, H., Tensho, A., et al. Lentysine: a new hypolipidemic agent from a mushroom. *Life Sci.* 1970; 9(7):379-385.

[59] Minamino, T., Yoshida, T., Tateno, K., et al. Ras induces vascular smooth muscle cell senescence and inflammation in human atherosclerosis. *Circulation.* 2003; 108(18): 2264-2269.

[60] Sugita, M., Sugita, H., Kaneki, M. Farnesyltransferase, monumycin a, prevents atherosclerosis development and reduces oxidative stress in apolipoprotein E-deficient mice. *Arterioscler. Thromb. Vasc. Biol.* 2007; 27(6):1390-1395.

[61] Michos, E. D., Melamed, M. L. Vitamin D and cardiovascular disease risk. *Curr. Opin. Clin. Nutr. Metab. Care.* 2008; 11(1):7-12.

[62] Sood, A., Arora, R. Vitamin D deficiency and its correlation with increased cardiovascular incidences. *Am. J. Ther.* 2010; 17(4):e105-109.

[63] Gouni-Berthold, I., Krane, W., Berthold, H. K. Vitamin D and cardiovascular disease. *Curr. Vasc. Pharmacol.* 2009; 7(3):414-422.

[64] Anagnostis, P., Athyros, V. G., Adamidou, F., et al. Vitamin D and cardiovascular disease: a novel agent for reducing cardiovascular risk? *Curr. Vasc. Pharmacol.* 2010; 8(5):720-730.

[65] Lee, J. H., O'Keefe, J. H., Bell, D., et al. Vitamin D deficiency: an important, common, and easily treatable cardiovascular risk factor? *J. Am. Coll. Cardiol.* 2008; 52(24):1949-1956.

[66] Mascitelli, L., Goldstein, M. R., Pezzetta, F. Vitamin D deficiency and cardiovascular disease. *Recenti. Prog. Med.* 2010; 101(5):202-211.

[67] Komoda, Y., Shimizu, M., Sonoda, Y., et al. Ganoderic acid and its derivatives as cholesterol synthesis inhibitors. *Chem. Pharmaceut. Bull.* 1989; 37(2): 531-533.

[68] J. Hajjaj H., Mace, C., Roberts, M., et al. Effect of 26-oxygenosterols from Ganoderma lucidum and their activity as cholesterol synthesis inhibitors. *Appl. Environ. Microbiol.* 2005; 71(7): 3653-3658.

[69] Agrawal, R. P., Chopra, G. S., Lavekar, M. M., et al. Effect of oyster mushroom on glycemia, lipid profile and quality of life in type 2 diabetic patients. *Australian J. Medical Herbalism.* 2010; 22(2):50-54.

[70] Valgimigli, M., Merli, E., Malagatti, P., et al. Endothelial dysfunction in acute and chronic coronary syndromes: evidence for a pathogenic role of oxidative stress. *Arch. Biochem. Biophys.* 2003; 420(2):255-261.

[71] Hansson, G. K. Inflammation, atherosclerosis, and coronary artery disease. *N Engl. J. Med.* 2005; 352:1685-1695.

[72] Martin, K. R. Both common and specialty mushrooms inhibit adhesion molecule expression and in-vitro binding of monocytes to human aortic endothelial cells in a pro-inflammatory environment. *Nutr. J.* 2010; 9(29):1-9.

[73] Martin, K. R. The bioactive agent ergothioneine, a key compound of dietary mushrooms, inhibits monocyte binding to endothelial cells characteristic of early cardiovascular disease. *J. Med. Food.* 2010; 13(6):1340-1346.

[74] Xu, C., HaiYan, Z., JianHang, Z., et al. The pharmacological effect of polysaccharides from Letinus edodes on the oxidative stress and expression of VCAM-1 mRNA of thoracic aorta endothelial cells in high-fat-diet rats. *Carbohydrate Poly.* 2008; 74:445-450.

[75] Percario, S., Odorizzi, V., Souza, D., et al. Edible mushroom Agaricus sylvaticus can prevent the onset of atheroma plaques in hypercholesterolemic rabbits. *Cell. Mol. Biol.* 2008; 54: 1055-1061.

[76] Barzilay, J. I., Abraham, L., Heckbert, S. R., et al. The relation of markers of inflammation to the development of glucose disorders in the elderly: the Cardiovascular Health Study. *Diabetes.* 2001; 50:2384-2389.

[77] Duncan, B. B., Schmidt, M. I., Pankow, J. S., et al. Low-grade systemic inflammation and the development of type 2 diabetes: the atherosclerosis risk in community study. *Diabetes.* 2003; 52:1799-1805.

[78] Pradhan, A. D., Manson, J. E., Rifai, N., et al. C-Reactive protein, interleukin 6, and risk of developing type 2 diabetes mellitus. *JAMA.* 2001; 286:327-334.

[79] Petersen, A. M., Pedersen, B. K. The anti-inflammatory effect of exercise. *J. Appl. Physiol.* 2005; 98:1154-1162.

[80] Hotamisligil, G. S. Inflammation and metabolic disorders. *Nature.* 2006; 444:860-867.

[81] Lee, Y. H., Pratley, R. E. The evolving role of inflammation in obesity and the metabolic syndrome. *Curr. Diabetes Rep.* 2005; 5:70-75.

[82] Plomgaard, P., Bouzakri, K., Krogh-Madsen, R., et al. Tumor necrosis factor-alpha induces skeletal muscle insulin resistance in healthy human subjects via inhibition of Akt substrate 160 phosphorylation. *Diabetes.* 2005; 54(10):2939-2945.

[83] Jedinak, A., Dudhgaonkar, S., Wu, Q. L., et al. Anti-inflammatory activity of edible oyster mushroom is mediated through the inhibition of NF-kB and AP-1. *Nutrition Journal.* 2011; 10(52):1-10.

[84] Lo, K. M., Cheung, P. C. K. Antioxidant activity of extracts from the fruiting bodies of Agrocybe aegerita var. alba. *Food Chem.* 2005; 89:533-539.

[85] Arduini, A., Eddy, L., Hochstein, P. The reduction of ferryl myoglobin by ergothioneine: a novel function for ergothioneine. *Arch. Biochem. Biophys.* 1990; 281(1):41-43.

[86] Brown, K. M., Arthur, J. R. Selenium, selenoproteins and human health: a review. *Public Health Nutr.* 2001; 4(2B):593-599.

[87] Werner, N., Kosiol, S., Schiegl, T., et al. Circulating endothelial progenitor cells and cardiovascular outcomes. *New Engl. J. Med.* 2005; 353:999-1007.

[88] Wannamethee, S. G., Shaper, A. G., Whincup, P. H., et al. Impact of diabetes on cardiovascular disease risk and all-cause mortality in older men: influence of age at onset, diabetes duration, and established novel risk factors. *Arch. Intern. Med.* 2011; 17(5):404-410.

[89] Reaven, G. M. Banting Lecture 1998. Role of insulin resistance in human disease. *Diabetes*. 1988; 37(12):1595-1607.

[90] Defronzo, R. A., Ferrannini, E. Insulin resistance: A multifaceted syndrome responsible for NIDDM, obesity, hypertension, dyslipidemia, and atherosclerotic cardiovascular disease. 1991; 14(3):173-194.

[91] Gray, A. M., Flatt, P. R. Insulin-releasing and insulin-like activity of Agaricus campestris (mushroom). *J. Endocrinol.* 1998; 157(2):259-266.

[92] Monchar, V., Talpur, N. A., Echard, B. W., et al. Effects of a water-soluble extract of maitake mushroom on circulating glucose/insulin concentrations in KK mice. *Diabetes, Obesity and Metabolism*. 2002; 4(1):43-48.

[93] Chobanian, A. V., Bakris, G. L., Black, H. R., et al. The Seventh Report of the Joint National Committee on Prevention, Detection, Evaluation, and Treatment of High Blood Pressure: The JNC 7 Report. *JAMA*. 2003; 289(19):2560-2572.

[94] Whelton, P. K., He, J., Appel, L. J., et al. Primary prevention of hypertension: clinical and public health advisory from The National High Blood Pressure Education Program. *JAMA*. 2002; 288(15):1882-1888.

In: Nutrigenomics
Editor: M. T. Ravi Subbiah

ISBN: 978-1-62417-051-5

Chapter V

Metabolic and Genetic Factors Responsible for Variable Inter-Individual Response of Blood Lipids to Nutraceutical Therapy

Todd C. Rideout*
Department of Exercise and Nutrition Sciences,
University of Buffalo, Buffalo, New York, US

Abstract

Nutraceuticals are increasingly recognized as effective modulators of lipid and lipoprotein metabolism and will likely have a prominent role in both the prevention and treatment of a wide range of dyslipidemic conditions in future health care systems. However, it is also recognized that the magnitude and direction of lipid responses to nutraceutical therapy is often not consistent on a patient-to-patient basis. These variable lipid responses are underscored by numerous metabolic, pathological, environmental, and genetic factors. The significance of nutraceuticals in future personalized lipid-lowering stratagems will depend greatly on our ability to apply advances in nutrigenomic technology to identify those subject-specific factors that influence how different individuals respond to specific nutraceutical therapies. This chapter will explore what is known regarding subject-specific lipid responses to nutraceuticals and discussadvances that have been made in identifying metabolic and genetic factors that may account for these variable responses.

Keywords: nutraceuticals; lipid lowering; variable response; personalized therapy; plant sterols; ω-3 fatty acids; soy isoflavones; coronary artery disease

* Corresponding author: Todd C. Rideout, Assistant Professor. Department of Exercise and Nutrition Sciences, University at Buffalo, 209A Kimball Tower, Phone: (716) 829-6786, E-mail: rideout@buffalo.edu.

Introduction

Recent assessment of death/disability and health care expenditures associated with coronary heart disease (CHD) demonstrate the tremendous impact that cardiovascular diseases, including acute myocardial infarction, ischemic heart disease, peripheral vascular disease, hypertension, and stroke, have on the lives of Americans. Conventional risk factors, including dyslipidemia, smoking, physical inactivity, obesity, diabetes, and hypertension contribute substantially to the incidence of CHD. The total annual direct and indirect costs associated with CHD in North America are estimated to be in excess of $400 billion. Unfortunately, the number of cardiovascular cases worldwide is expected to grow over the next 20 years.

Much of the overwhelming incidence of cardiovascular disease in modern societycan be linked with the rapid evolution of human dietary practices. Prior to the agricultural revolution, diets of our Paleolithic ancestors drew heavily upon a variety of food sources including lean game, seafood, and fruits and vegetables, and were therefore rich in fibers, phytochemicals, and animal and vegetable proteins. Conversely, the industrial revolution changed daily living habits to such an extent that pre-packaged, ready-to-eat foods became a necessary part of typical eating behavior. These 'industrial' diets were typified by foods that were highly processed and rich in simple carbohydrates, saturated fats, and cholesterol. Today, with an understanding of the role of diet in disease progression and a renewed public conscience toward healthy eating,daily food habits trend toward a re-introduction of dietary bioactive health-promoting components. The global functional food and nutraceutical market is steadily growing, driven by high expectations from industry stakeholders with increasing market potential and informed consumers demanding healthy and therapeutic food options. Consumer data suggest an increase in the awareness and use of functional foods and nutraceuticals amongst Americans [1]. Furthermore, with the availability of evidenced-based data from controlled intervention studies, cardiologists are increasingly recommending cholesterol-lowering nutraceuticals for at-risk patients [2]. Although modern society may be more aware of the role of diet and specific dietary components in disease prevention and treatment, there is still an expectation for highly convenient, practical dietary choices that fit into today's hectic lifestyle. While the availability of functional foods offer increased convenience and functionality over strict dietary changes, the delivery of purified bioactives through pills, capsules, and supplements may be viewed as even more desirable in a society that has become accustomed to 'pill-popping'. However, debate continues around the health implications of nutrients in whole-foods versus nutraceutical delivery and the most appropriate dietary recommendations that re-enforce the value of a balanced diet in addition to nutraceutical supplementation for specific health-promoting purposes.

Lipid-Lowering Response to Nutraceuticals

Dietary bioactive compounds, whether delivered as a purified pill/capsule isolate or formulated for food matrix incorporation, are increasingly recognized as effective therapeutic options for dyslipidemia. Dietary bioactives compounds that have been extensively investigated for their lipid-modulating properties in both pre-clinical and clinical studies include, cholesterol, soluble fibers, plant sterols, calcium, probiotics, and omega 3-fatty acids. Although lipid reductions in response to nutraceutical consumption may not be in the same magnitude as with pharmaceutical interventions, even comparatively modest improvements in CVD risk markers may have significant impacts on public health and health care expenditures if applied consistently to a large population over a long-term basis [3].

Similar to modern pharmaceutical therapies, dietary bioactive compounds affectblood lipoprotein metabolism by modulating the expression of intracellular lipid homeostatic regulators at the DNA, RNA and protein levels to transform lipid phenotypes at a tissue and organ level [4]. Although diet and drug-based lipid-lowering therapies share similar molecular mechanisms,they also share another important attribute that can often limit their clinical utility - variable inter-individual responses [5, 6]. A review of the clinical trial database clearly demonstrates considerable inter-individual heterogeneity in the responsiveness of plasma lipids to nutraceutical therapies [7]. A prime example of this inter-individual response is demonstrated from a recent intervention study following 43 participants consuming a low fat diet with added soy protein (25g) and dietary fiber (15g) for 1 year [8]. When subjects were stratified as hyper-responders (HYPER), normal responders (NR), or hypo-responders (HYPO) according to their percent change in blood lipids, a great range of responsiveness was observed for total cholesterol (51.2%HYPER, 25.5%NR; 23.3%HYPO), LDL-C (48.8% HYPER; 18.6% NR; and 32.6% HYPO), and TAG (76.8% HYPER; 11.6% NR, and 11.6% HYPO). Although dietary bioactive components elicit specific physiological responses, the magnitude and direction of these responses are influenced by multiple factors that can be broadly grouped into four categories: 1) physiological factors including readily definable characteristics such as ethnicity, age, and gender and less readily characterized metabolic markers (e.g. glucose tolerance); 2) disease associated factors including diabetes, metabolic syndrome, and dyslipidemia; 3) environmental factors including lifestyle and diet; and 4) genetic factors,the most common of which are single nucleotide polymorphisms (SNPs)(5). Characterizing these specific factors that influence individual response to nutraceutical therapy is considered a critical milestone in transitioning from conventional 'one-size fits all' nutritional guidelines to personalized dietary interventions tailored to the individual [9]. Recent work in identifying factors of nutraceutical response have focused heavily on genetics, however, moving forward with personalized nutrition therapy in the short-term will require the availability of readily accessible clinical markers of responsiveness [10]. Although assessment of responsiveness with genetic markers is foreseeable inclinics of the future, we are not yet at a stage where these markers are useful in current clinical practice. Therefore, current personalized nutrition therapy will most benefit fromidentifying response to nutraceuticals in readily identifiable patient subgroups [11].

Types of Variability

Before we examine variability of lipids to specific nutraceutical therapies, it may be useful to first define the types of variability that are encountered in the literature. One type of variability is the blood lipid responses that are reported between research investigations for any given nutraceutical therapy.

These inter-study inconsistencies are not routed in true metabolic responsesbut are instead a product ofunderlying differences in experimental design between research investigations includingthe study population, diet background, treatment dose and duration, statistical model employed, and phlebotomy protocol. In the case of the later, even seemingly trivial aspects of the phlebotomy procedure, including posture, may contribute to day-to-day fluctuations in plasma cholesterol and underlie much of the variability observed between research investigations [12]. The second type of variability is that observed within-a specific study population. This inter-individual heterogeneity of lipid responsiveness is most readily observed in randomized cross-over studieswhere the research subjects undergo both treatment and control phases and consequently serve as their own control [10]. This type of variation is most likely a result of true metabolic and genetic factors that underscore subject-specific response. Based on these response observations, three types of individuals can be identified according to his or herindividual response to nutraceutical intervention: 1) 'responders' who have clinically significant reductions in plasmalipidsfollowing a nutraceutical intervention; 2) 'non-responders' who exhibit no response or a sub-optimal plasma lipid response to a nutraceutical intervention; and 3) 'adverse responders' whoseplasma lipids increase following a nutraceutical intervention [13]. Examples of response, non-response, and adverse response have been clearly established following nutraceutical (plant sterol)and drug (statin)-basedlipid lowering treatments [14, 15].

Variable Lipid Responses to Specific Dietary Bioactives

Cholesterol

Perhaps the most misunderstood and controversial relationship between a dietary bioactive compound and a blood lipid response concerns the influence of dietary cholesterol on plasma cholesterol concentrations. It is quite clear from clinical cholesterol feeding studies that individuals do not display a uniform and consistent plasma lipid response to dietary cholesterol consumption [16]. Although the connection between blood cholesterol and atherosclerosis is well established, the perceived effect of dietary cholesterol on blood cholesterol concentrations has been over-emphasized, even in the face of a significant body of research suggesting a mainly nominal relationship [17, 18]. Animal model studies initially provided evidence suggesting the existence of responders and non-responders to dietary cholesterol [19]. Selection studies in the 1970s using the Sprague-Dawley rat suggested that hypo- and hyper-response to dietary cholesterol had a genetic basis [20]. Furthermore, due to enhanced VLDL production and reduced hepatic LDL plasma clearance, pregnant versus virgin rats were shown to have a greater lipidemic response to high a cholesterol diet [21].

Clinical studies soon followed that sought to determine if responders/non-responders to dietary cholesterol could be identified in a human study population and to quantify the inter-individual variance in cholesterol response [19]. It has since been suggestedthat 1 out of every 3 individuals can be classified as a hyper-responder to dietary cholesterol [22]. Although hyper-responders display increased plasma LDL-C following dietary cholesterol intake, these subjects also exhibit augmentedactivity of plasma enzymes that regulate lipoprotein metabolism (lecithin cholesterol acyltransferase and cholesteryl ester transfer protein), alterations in cholesterol synthesis, a rise in HDL-C, and an increase in LDL particle size, all of which may enhance reverse cholesterol transport and reduce heart disease risk [22-24].

The lipoprotein response to dietary cholesterol can be affected by hormonal status, ethnicity, age, and genetic predisposition. At the genetic level, previous work suggests that subjects possessing a C/C genotype in the polymorphic site of codon 640 region of ABCG5 display a greater increase in plasma LDL-C afterconsuming 640 mg/d of cholesterol for 30 days in the form of eggs compared with those individuals with a G/G or G/C genotype [16]. Alternatively, the same study population was examined for associations between polymorphisms in apoC3 and apoA4 and the plasma cholesterol response to egg consumption, however, no associations to dietary cholesterol could be found [25].

Soy Protein and Isoflavones

Based largely on the results of a meta-analysis in the mid-1990s [26], consumption of soy is supported by the US Food and Drug Administration and the American Heart Association for the reduction of blood total and LDL-C. However, in recent years the initial enthusiasm surrounding the lipid-lowering effects of soy has waned in the face of numerous studies reporting either no response [27, 28] or variable LDL-C lowering response in the range of 5-28% [8, 29]. Another controversy is whether the reputed lipid-lowering effects of soy are directly related to the soy protein or the isoflavone component. Some research suggests that the health benefits of soy are evident only after large intestinal bacterial conversion of the isoflavone component to equol, a bacterial metabolite with heightened estrogenic activity.In fact, it has been suggested that much of the variable lipid-lowering response to soy is related to inter-individual differences in the efficacy of equol production as only 30-50% of the population have the intestinal bacterial to produce equol from daidzein, a primary soy-derived isoflavone [30]. However, stratification of the lipid-response to soy consumption based on equol production status has met with inconsistent results. While observing no lipid response in a hypercholesterolemic/hypertensive study population following soy treatment for 5 weeks, Meyer et al. reported lower plasma cholesterol and TAG in a sub-set of 8 equol positive [31]. Alternatively, equol production status did not appear to be a reliable indicator of lipid-response to soy consumption in hypercholesterolemic women fed a soy protein isolate with or without a probiotic [32].

A previous meta-analysis examining lipid response to soy isoflavones suggests that reductions in total and LDL-C are more evident in hypercholesterolemic individuals compared with subjects with fasting baseline cholesterol in the normal range [33]. Results from a soy intervention study that prospectively recruited hyperlipidemic and normolipidemic hemodialysis patients also reported significant lipid reductions in the hyper vs. the normolipidemic groups following consumption of 30 g of isolated soy protein for 12 weeks [34].

The effect of genetic variation on the lipid-lowering response to soy has been the focus of few research investigations. Although no difference in plasma lipids were observed versus the placebo group in 177 women fed 43.5 mg of isoflavones/d for 1 year, a retrospective examination of lipids associated with apoE genotype revealed a greater total and LDL-C lowering response in E2/E3 women compared with E3/E3 or E3/E4 subjects [35].

However, in direct contrast to this observation, Gaddi et al. observed greater lipid reductions in apoE3/E3 and apoE3/E4 compared with a negligible response in E3/E2 male and female subjects consuming soy protein for 4 weeks [36]. Sex differences may partially explain the distinct outcomes between these two studies; however, previous work suggests no difference in the lipid-lowering response to soy consumption between men and women [37].

Finally, polymorphisms in the estrogen receptor β gene have been shown to affect the lipid-response to dietary isoflavones. Healthy post-menopausal women genotyped as AA in the estrogen receptor β gene (TSP5091) have been shown to respond with increases in HDL-C more favorable than GG or GA subjects following consumption of isoflavone-enriched bars for 8 weeks [38].

Dietary Fiber

The viscous forming, soluble fibers including pectin, psyllium, and guar gum have a long history as effective LDL-cholesterol lowering agents in both animal and human clinical investigations. Soluble fibers elicit their hypocholesterolemic effects within the intestine by interfering with the enterohepatic circulation of cholesterol and bile acids and promoting bacterial fermentation in the large intestinewith the production of short-chain fatty acids [39].

Secondary hepatic responses have also been described with modulation in the expression and activity of hepatic regulators of cholesterol synthesis and bile acid conversion, VLDL packaging and secretion, and plasma LDL-C clearance. Total cholesterol reductions in response to soluble fiber consumption vary considerably ranging from -17% to 3% for psyllium, from -16% to -5% for pectin, and from -17% to 4% for guar gum [40]. Part of this variability, particularly between fibers,may be due to the fact that soluble fibers differ in their physiochemical properties and therefore elicit unique physiological responses with different magnitudes of cholesterol lowering. Interestingly, cecal SCFA production, particularly butyrate, in response to dietary fiber consumption greatly varies between individuals and may also explain much of the heterogeneity of health responses to dietary fiber [41].

Similar to soy, dietary fibers appear to exert the greatest cholesterol-lowering benefit in hypercholesterolemic individuals compared with normocholesterolemic subjects [42, 43].

Studies comparing the lipid-lowering efficacy of soluble fibers between men and women suggest a definite sex disparity of response, however, results of gender effects are often contradictory. Consumption of both flaxseed [43] and a viscous fiber blend of konjac, sodium alginate and xanthan gum [44] have been reported to have greater cholesterol-lowering efficacy in women than in men.

Furthermore, in a cross-section analysis of type 1 diabetics in >1000 male and female subjects, intake of total, soluble, and insoluble fiber were inversely associated with CVD in women but not in men [45].

Among women, post-menopausal women appear to be more responsive to the lipid modulating effects of dietary fibers compared with pre-menopausal women [43, 46]. Augmented lipid-lowering responses to dietary fibers in post-menopausal women may be due to increased dyslipidemia that is observed with the loss of estrogen function.

Alternatively, other dietary fiber intervention studies have observed greater response in men compared with women [3, 47].

Given the action of dietary fibers within the intestine, it would seemappropriate to examine variable responses to fiber in the context of polymorphic intestinal candidate genes that regulate fat and cholesterol digestion/absorption, however,relatively little work has been done in this area. One candidate gene that has been examined is fatty acid binding protein 2 (FABP2), a polymorphic gene (codon 54) whose protein product binds both unsaturated and saturated long-chain fatty acids within the enterocyte. When 43 subjects were fed soluble vs. insoluble fiber in a cross-over design, those homozygous for FABP2 T54 had greater decreases in plasma total and LDL-C, and apoB compared with A54 subjects [48]. The T54 variant of FABP2 has also been shown to be associated with higher HDL-C concentrations and apo A-IV concentrations following consumption of soluble fiber (*P. ovate* husk) compared with the insoluble fraction (*P. ovate* seeds)in 28 men diagnosed with CVD [49]. The TAG-lowering response to *P. ovate* husk has also been shown to be influenced by variants in FABP2 (rs1799883) and plasminogen-activator-inhibitor (rs1799768) in hypercholesterolemic patients compared with cellulose [50].

Polymorphisms in 3-hydroxy-3-methyl-glutaryl-CoA (HMG-CoA) reductase, the rate-limiting enzyme in cholesterol biosynthesis, have also been associated with lipid responses to dietary fiber consumption.

In a retrospective cross-sectional study of 2300 participants, carriers of the G allele of HMG-CoAr (rs17238540) were more responsive to the TAG lowering actions of dietary fiber compared with TT individuals [51].

Genetic variants in apoE may also underscore subject specific differences in blood lipids following oat and wheat bran consumption. In a short-term feeding study (2 weeks) in 67 subjects, LDL-C in E2 carriers was more responsive to fiber consumption compared with E3 homozygotes and E4 carriers [52].

Omega 3 Fatty Acids

Consumption of omega 3 (*n*-3) fatty acids, including marine-derived eicosapentaenoic (20:5n-3, EPA) and docosahexaenoic acid (22:6n-3, DHA) and their plant-based precursor alpha-linolenic acid (18:3*n*-3, ALA) have been intensively investigated for the their lipid-lowering properties.

Although EPA and DHA are well-characterized TAG-lowering agents, the TAG-lowering response to ALA is considered more variable, most likely due to inter-individual differences in theefficiency of ALA conversion to EPA (8-20%) and DHA (0.9-5%) through the FADS1/2 gene cluster [53]. Gender studies suggest that women have a 2.5 fold higher efficacy of ALA to EPA conversion compared with men. Single nucleotide polymorphisms in FADS1, FADS2, and FADS3 have been shown to effect plasma and tissue DHA concentrations during pregnancy [54].

Along with TAG-reductions, consumption of marine-derived omega 3 fatty acids often result in an increase in LDL-C in the range of 5-10% that is characterized by a significant degree of inter-individual variability [9, 55]. The LDL-C response to 3g of EPA + DHA has been reported to be from -49 to 87% [56]. Combining EPA/DHA with a statin has been suggested as a potentially effective combination therapy to offset potential increases in LDL-C with n-3 fatty acid consumption.

Retrospective data analysis from the COMBination study suggests that patients with low baseline LDL-C (while on statin monotherapy) respond more favorably to an *n*-3 fatty acid/simvastatin combination with the largest reductions in plasma TAG concentrations and less pronounced increase in LDL-C compared to patients with high baseline LDL-cholesterol [57]. Amongst nutraceuticals, omega-3 fatty acids have been studied most extensively with respect to the association between genetic variants and lipid response. Results of several investigations suggest that apoE4 carriers may be more susceptible to the LDL-C raising effects of n-3 fatty acids.

In a prospective recruitment study of apo E3/E3 or E3/E4 subjects fed an EPA rich oil (3.3 g/d), a DHA rich oil (3.7 g/d) or a control oil in a 4-week crossover study, Onlano-Martin et al. (2010) [58] reported that the LDL-raising effect of fish oil was largely confined to apo E4 carriers following DHA consumption.

Furthermore, compared with common E3/E3 individuals, apoE4 subjects responded with a higher increase in total cholesterol in male subjects fed 3 g EPA/DHA compared with olive oil [59]. If confirmed within a larger population, this data may result in future personalized nutrition guidelines recommendingfish oil for TAG-lowering in apoE3 patients.

Omega-3 fatty acids have been speculated to modulate tissue and plasma lipid homeostasis by serving as ligands for the PPAR nuclear receptor. Accordingly, the Pro12ALA polymorphism of PPAR-γ has been shown to influence a broad range of lipidresponses to EPA/DHA consumption. Carriers of the ALA12 allele have been shown to respond with greater TAG reductions following consumption of 2.4 g/d of EPA/DHA for 3 months [60]. Furthermore, a previous cross-sectional analysis suggests that the Pro12 allele may underscore fasting free fatty acid concentrations in response to DHA consumption [61].

Polymorphisms in intestinal FABP-2have only recently been investigated as a potential contributor to the lipoprotein response to fish oil consumption. Duodenal FABP2 expression has recently been shown to be dependent on n-3 fatty acid intake in homozygotes for haplotype A compared with B carriers of the FABP2 promoter [62].

Furthermore, hypertriglyceridemic Thr54 carriers have been shown to demonstrate greater responsiveness in plasma TAG, VLDL-cholesterol, and HDL-cholesterol following consumption of 2 g of EPA for 2 weeks compared with ALA54 carriers [63].

Nitric oxide synthase (NOS), the key gene regulating nitric oxide production for endothelial function, has also been shown to be involved in regulating circulating cholesterol and TAG concentrations [64] and is responsive to fish oil feeding [65]. Recently, Fergusen et al. (2010) observed a significant interaction between the NOS3 polymorphism (rs1799983) and plasma TAG concentrations with minor allele carriers displaying increased sensitivity to fish oil consumption [66].

Plant Sterols

Plant sterols, constituents found in plants that chemically resemble cholesterol, are arguably the best-defined nutraceutical approach to reducing blood cholesterol concentrations. Consumption of plant sterols at the recommended dose of 2 g per day has been demonstrated to consistently reduce circulating LDL-cholesterol levels in humans by approximately 10% with little or no effect on HDL-C [67]. North Americans consume plant sterol at a level of around 300-400 g/d, mainly from food constituents such as vegetable oils, nuts, seeds, grain products, fruits and vegetables. It is generally accepted that plant sterols lower blood cholesterol concentrations by interfering intestinal cholesterol absorption by competing with cholesterol of dietary and biliary origin for micellular incorporation [68].

Despite repeated demonstrations of the cholesterol-lowering efficacy of plant sterols, clinical trials clearly demonstrate that not all individuals respond equally to standard recommended plant sterol dosages. Much of this variable cholesterol-lowering response to plant sterol therapy between published reports is likely related to differences in experimental design such as food matrixing and dosing, but there has been increasing focus on the role of metabolic andgenetic factorsthat may underscore subject specific responses [69]. Recent work suggests that basal cholesterol synthesis rate may predict an individual's cholesterol-lowering response to plant sterols. In a retrospective analysis of 3 controlled feeding PS interventions employing stable isotopes to directly measure fractional cholesterol synthesis rate (FSR), basal cholesterol FSR was 23% higher in the non-responder versus responder subgroups, suggesting that patients with high basal cholesterol synthesis are unresponsive to plant sterol treatment [70]. Genetic response studies with respect to plant sterols have focused on genes that regulate lipoprotein transport, hepatic cholesterol synthesis, and enterohepatic sterol circulation, however it is difficult to come to any consensus as study results are often contradictory. Two previous intervention studies have reported that subjects with an apoE4 allele may be more responsive to the LDL-C lowering effects of plant sterols compared with individuals homozygous for the apoE3 allele [71, 72]. However, other reports suggest that apoE genotype does not predict individual lipid responses to plant sterol consumption [73-75]. Only two studies have examinedplant sterol responsiveness in association withNiemann-Pick C1 Like 1 variants with contradictory results. While Zhao et al. (2008) suggest that mutant alleles of the NPC1L1 haplotype of 872 C>G (L272L) and 3929 G>A (Y1291Y) have greater LDL-C reductions to plant sterols compared with wild type counterparts [76], Rudkowska et al. (2008) reported no association between the same NPC1L1 variants and cholesterol responsiveness [77]. Similarly, while the majority of previous reports were unable to detect significant associations between genetic variants in the sterol-efflux proteins ATP-binding cassette transporters ABCG5 and G8 [77-79], carriers of the A allele of the T400 K ABCG8 polymorphism were reported to have a greater LDL-C lowering response to plant sterol consumption in one study [76].

Conclusion

From the above discussion it is apparent that while nutraceuticals can be effective lipid-lowering agents, determining metabolic and genetic factors that influence the broad range of lipid responses between individuals must be a research priority. Although nutrigenomics makes it possible to identify the genetic basis of this variability, information on the influence of genetic polymorphisms on response to nutraceuticalsare, in many cases, lagging behind response studies to whole dietary patterns such as carbohydrate and fat load. There are numerous nutraceuticals with variable lipid-lowering activity that have yet to be investigated at the genomic level including probiotics [80], calcium [81], alpha-lipoic acid [82], garlic [83], and polycosinols [84].

The lack of consistency between the results of nutraceutical responsiveness studies will also demand research attention. The majority of recent responsiveness investigations are associative studies that retrospectively correlate physiological endpoints with individual lipid responsesto anutraceutical intervention [9]. However, controlled prospective studies that actively recruit by specific physiological, disease state or genetic variables are needed to identify true metabolic and genetic factors of nutraceutical response. Another potential reason for the lack of consistency observed between studies examining the role of genetic polymorphisms and response to nutraceuticals is that most studies focus on a single gene variant, whereas in reality, subject specific response to any dietary intervention is under polygenic control.

Overall the future of nutraceuticals in the prevention and treatment of dyslipidemias is encouraging and holds practical clinical promise. The identification of genetic traits that confer variability will eventually be used in conjunction with rapid genotyping assays to allow future practitioners to determine thespecific nutraceutical intervention that will be most effective in at-risk patients.

References

[1] International Food Information Council: Functional Foods/ Foods for Health Consumer Trending Survey: Executive Research Report (2009). Available from: http://www.foodinsight.org/Resources/Detail.aspx?topic=2009_Functional_Foods_Foods_For_Health_Consumer_Trending_Survey_Executive_Summary.

[2] Dickinson, A., Shao, A., Boyon, N., Franco, J. C. Use of dietary supplements by cardiologists, dermatologists and orthopedists: report of a survey. *Nutr. J.* 2011;10:20.

[3] Davidson, M. H., Maki, K. C., Kong, J. C., Dugan, L. D., Torri, S. A., Hall, H. A., et al. Long-term effects of consuming foods containing psyllium seed husk on serum lipids in subjects with hypercholesterolemia. *Am. J. Clin. Nutr.* 1998;67:367-76.

[4] Kussmann, M., Affolter, M. Proteomics at the center of nutrigenomics: comprehensive molecular understanding of dietary health effects. *Nutrition*. 2009;25:1085-93.

[5] Rideout, T. C. Getting personal: considering variable interindividual responsiveness to dietary lipid-lowering therapies. *Curr. Opin. Lipidol.* 2011;22:37-42.

[6] Romaine, S. P., Bailey, K. M., Hall, A. S., Balmforth, A. J. The influence of SLCO1B1 (OATP1B1) gene polymorphisms on response to statin therapy. *Pharmacogenomics J.* 2010;10:1-11.

[7] Lairon, D., Defoort, C., Martin, J. C., Amiot-Carlin, M. J., Gastaldi, M., Planells, R. Nutrigenetics: links between genetic background and response to Mediterranean-type diets. *Public Health Nutr.* 2009;12:1601-6.

[8] Torres, N., Guevara-Cruz, M., Granados, J., Vargas-Alarcon, G., Gonzalez-Palacios, B., Ramos-Barragan, V. E., et al. Reduction of serum lipids by soy protein and soluble fiber is not associated with the ABCG5/G8, apolipoprotein E, and apolipoprotein A1 polymorphisms in a group of hyperlipidemic Mexican subjects. *Nutr. Res.* 2009;29:728-35.

[9] Lovegrove, J. A., Gitau, R. Personalized nutrition for the prevention of cardiovascular disease: a future perspective. *J. Hum. Nutr. Diet.* 2008;21:306-16.

[10] Zivkovic, A. M., Wiest, M. M., Nguyen, U., Nording, M. L., Watkins, S. M., German, J. B. Assessing individual metabolic responsiveness to a lipid challenge using a targeted metabolomic approach. *Metabolomics*. 2009;5:209-18.

[11] Ordovas, J. M. Genotype-phenotype associations: modulation by diet and obesity. *Obesity* (Silver Spring). 2008;16 Suppl. 3:S40-6.

[12] Dixon, M., Paterson, C. R. Posture and the composition of plasma. *Clin. Chem.* 1978;24:824-6.

[13] Bach, R. G. Heterogeneity of response to lipid-lowering therapy. *J. Am. Coll. Cardiol.* 2001;38:2136-7.

[14] Rideout, T. C., Chan, Y. M., Harding, S. V., Jones, P. J. Low and moderate-fat plant sterol fortified soymilk in modulation of plasma lipids and cholesterol kinetics in subjects with normal to high cholesterol concentrations: report on two randomized crossover studies. *Lipids Health Dis.* 2009;8:45.

[15] Mangravite, L. M., Wilke, R. A., Zhang, J., Krauss, R. M. Pharmacogenomics of statin response. *Curr. Opin. Mol. Ther.* 2008; 10:555-61.

[16] Herron, K. L., McGrane, M. M., Waters, D., Lofgren, I. E., Clark, R. M., Ordovas, J. M., et al. The ABCG5 polymorphism contributes to individual responses to dietary cholesterol and carotenoids in eggs. *J. Nutr.* 2006;136:1161-5.

[17] Fernandez, M. L., Calle, M. Revisiting dietary cholesterol recommendations: does the evidence support a limit of 300 mg/d? *Curr. Atheroscler. Rep.* 2010;12:377-83.

[18] Kratz, M. Dietary cholesterol, atherosclerosis and coronary heart disease. *Handb. Exp. Pharmacol.* 2005(170):195-213.

[19] Katan, M. B., Beynen, A. C., de Vries, J. H., Nobels, A. Existence of consistent hypo- and hyperresponders to dietary cholesterol in man. *Am. J. Epidemiol.* 1986;123:221-34.

[20] Imai, Y., Matsumura, H. Genetic studies on induced and spontaneous hypercholesterolemia in rats. *Atherosclerosis*. 1973;18:59-64.

[21] Munilla, M. A., Herrera, E. A cholesterol-rich diet causes a greater hypercholesterolemic response in pregnant than in nonpregnant rats and does not modify fetal lipoprotein profile. *J. Nutr.* 1997;127:2239-45.

[22] Greene, C. M., Waters, D., Clark, R. M., Contois, J. H., Fernandez, M. L. Plasma LDL and HDL characteristics and carotenoid content are positively influenced by egg consumption in an elderly population. *Nutr. Metab.* (Lond.). 2006;3:6.

[23] Herron, K. L., Vega-Lopez, S., Conde, K., Ramjiganesh, T., Shachter, N. S., Fernandez, M. L. Men classified as hypo- or hyperresponders to dietary cholesterol feeding exhibit differences in lipoprotein metabolism. *J. Nutr.* 2003;133:1036-42.

[24] Mistry, P., Miller, N. E., Laker, M., Hazzard, W. R., Lewis, B. Individual variation in the effects of dietary cholesterol on plasma lipoproteins and cellular cholesterol homeostasis in man. Studies of low density lipoprotein receptor activity and 3-hydroxy-3-methylglutaryl coenzyme A reductase activity in blood mononuclear cells. *J. Clin. Invest.* 1981 ;67:493-502.

[25] Herron, K. L., Lofgren, I. E., Adiconis, X., Ordovas, J. M., Fernandez, M. L. Associations between plasma lipid parameters and APOC3 and APOA4 genotypes in a healthy population are independent of dietary cholesterol intake. *Atherosclerosis*. 2006;184:113-20.

[26] Anderson, J. W., Johnstone, B. M., Cook-Newell, M. E. Meta-analysis of the effects of soy protein intake on serum lipids. *N Engl. J. Med.* 1995;333:276-82.

[27] Sacks, F. M., Lichtenstein, A., Van Horn, L., Harris, W., Kris-Etherton, P., Winston, M. Soy protein, isoflavones, and cardiovascular health: a summary of a statement for professionals from the american heart association nutrition committee. *Arterioscler. Thromb. Vasc. Biol.* 2006;26:1689-92.

[28] Liu, Z. M., Ho, S. C., Chen, Y. M., Ho, Y. P. The effects of isoflavones combined with soy protein on lipid profiles, C-reactive protein and cardiovascular risk among postmenopausal Chinese women. *Nutr. Metab. Cardiovasc. Dis.* 2011 Mar. 21.

[29] Zhan, S., Ho, S. C. Meta-analysis of the effects of soy protein containing isoflavones on the lipid profile. *Am. J. Clin. Nutr.* 2005;81:397-408.

[30] Ko, T. F., Tsai, H. S., Lin, S. M., Liu, C. D., Learn, S. P., Chiou, R. Y. GC-MS Determined Distribution of Urinary Equol Producers as Affected by Age, Gender, and Repeated Ingestions of Soymilk. *J. Food Sci.* 2010;75:H306-10.

[31] Meyer, B. J., Larkin, T. A., Owen, A. J., Astheimer, L. B., Tapsell, L. C., Howe, P. R. Limited lipid-lowering effects of regular consumption of whole soybean foods. *Ann. Nutr. Metab.* 2004;48:67-78.

[32] Greany, K. A., Nettleton, J. A., Wangen, K. E., Thomas, W., Kurzer, M. S. Probiotic consumption does not enhance the cholesterol-lowering effect of soy in postmenopausal women. *J. Nutr.* 2004;134:3277-83.

[33] Taku, K., Umegaki, K., Sato, Y., Taki, Y., Endoh, K., Watanabe, S. Soy isoflavones lower serum total and LDL cholesterol in humans: a meta-analysis of 11 randomized controlled trials. *Am. J. Clin. Nutr.* 2007;85:1148-56.

[34] Chen, S. T., Ferng, S. H., Yang, C. S., Peng, S. J., Lee, H. R., Chen, J. R. Variable effects of soy protein on plasma lipids in hyperlipidemic and normolipidemic hemodialysis patients. *Am. J. Kidney Dis.* 2005;46:1099-106.

[35] Atkinson, C., Oosthuizen, W., Scollen, S., Loktionov, A., Day, N. E., Bingham, S. A. Modest protective effects of isoflavones from a red clover-derived dietary supplement on cardiovascular disease risk factors in perimenopausal women, and evidence of an interaction with ApoE genotype in 49-65 year-old women. *J. Nutr.* 2004;134:1759-64.

[36] Gaddi, A., Ciarrocchi, A., Matteucci, A., Rimondi, S., Ravaglia, G., Descovich, G. C., et al. Dietary treatment for familial hypercholesterolemia--differential effects of dietary soy protein according to the apolipoprotein E phenotypes. *Am. J. Clin. Nutr.* 1991;53:1191-6.

[37] Teede, H. J., Dalais, F. S., Kotsopoulos, D., Liang, Y. L., Davis, S., McGrath, B. P. Dietary soy has both beneficial and potentially adverse cardiovascular effects: a placebo-controlled study in men and postmenopausal women. *J. Clin. Endocrinol. Metab.* 2001;86:3053-60.

[38] Hall, W. L., Vafeiadou, K., Hallund, J., Bugel, S., Reimann, M., Koebnick, C., et al. Soy-isoflavone-enriched foods and markers of lipid and glucose metabolism in postmenopausal women: interactions with genotype and equol production. *Am. J. Clin. Nutr.* 2006;83:592-600.

[39] Rideout, T., Harding, S., Mz, F., Jones, P. Guar gum and similar soluble fibers in the regulation of cholesterol metabolism: Current understandings and future research priorities. *Vascular Health and Risk Management*. 2008;4:1023-33.

[40] Brown, L., Rosner, B., Willett, W. W., Sacks, F. M. Cholesterol-lowering effects of dietary fiber: a meta-analysis. *Am. J. Clin. Nutr.* 1999;69:30-42.

[41] McOrist, A. L., Miller, R. B., Bird, A. R., Keogh, J. B., Noakes, M., Topping, D. L., et al. Fecal butyrate levels vary widely among individuals but are usually increased by a diet high in resistant starch. *J. Nutr.* 2011;141:883-9.

[42] Ripsin, C. M., Keenan, J. M., Jacobs, D. R., Jr., Elmer, P. J., Welch, R. R., Van Horn, L., et al. Oat products and lipid lowering. A meta-analysis. *JAMA*. 1992;267:3317-25.

[43] Pan, A., Yu, D., Demark-Wahnefried, W., Franco, O. H., Lin, X. Meta-analysis of the effects of flaxseed interventions on blood lipids. *Am. J. Clin. Nutr.* 2009;90:288-97.

[44] Lyon, M., Wood, S., Pelletier, X., Donazzolo, Y., Gahler, R., Bellisle, F. Effects of a 3-month supplementation with a novel soluble highly viscous polysaccharide on anthropometry and blood lipids in nondieting overweight or obese adults. *J. Hum. Nutr. Diet*. 2011;24:351-9.

[45] Toeller, M., Buyken, A. E., Heitkamp, G., de Pergola, G., Giorgino, F., Fuller, J. H. Fiber intake, serum cholesterol levels, and cardiovascular disease in European individuals with type 1 diabetes. EURODIAB IDDM Complications Study Group. *Diabetes Care*. 1999; 22 Suppl. 2:B21-8.

[46] Ganji, V., Kuo, J. Serum lipid responses to psyllium fiber: differences between pre- and post-menopausal, hypercholesterolemic women. *Nutr. J.* 2008;7:22.

[47] Jenkins, D. J., Wolever, T. M., Rao, A. V., Hegele, R. A., Mitchell, S. J., Ransom, T. P., et al. Effect on blood lipids of very high intakes of fiber in diets low in saturated fat and cholesterol. *N Engl. J. Med.* 1993;329:21-6.

[48] Hegele, R. A., Wolever, T. M., Story, J. A., Connelly, P. W., Jenkins, D. J. Intestinal fatty acid-binding protein variation associated with variation in the response of plasma lipoproteins to dietary fibre. *Eur. J. Clin. Invest.* 1997;27:857-62.

[49] Sola, R., Godas, G., Ribalta, J., Vallve, J. C., Girona, J., Anguera, A., et al. Effects of soluble fiber (Plantago ovata husk) on plasma lipids, lipoproteins, and apolipoproteins in men with ischemic heart disease. *Am. J. Clin. Nutr.* 2007;85:1157-63.

[50] Sola, R., Bruckert, E., Valls, R. M., Narejos, S., Luque, X., Castro-Cabezas, M., et al. Soluble fibre (Plantago ovata husk) reduces plasma low-density lipoprotein (LDL) cholesterol, triglycerides, insulin, oxidised LDL and systolic blood pressure in hypercholesterolaemic patients: A randomised trial. *Atherosclerosis*. 2010;211:630-7.

[51] Freitas, R. N., Khaw, K. T., Wu, K., Bowman, R., Jeffery, H., Luben, R., et al. A single nucleotide polymorphism in the 3-hydroxy-3-methylglutaryl-coenzyme A reductase gene (HMGCR) influences the serum triacylglycerol relationship with dietary fat and fibre in the European Prospective Investigation into Cancer and Nutrition in Norfolk (EPIC-Norfolk) study. *Br. J. Nutr.* 2010;104:765-72.

[52] Jenkins, D. J., Hegele, R. A., Jenkins, A. L., Connelly, P. W., Hallak, K., Bracci, P., et al. The apolipoprotein E gene and the serum low-density lipoprotein cholesterol response to dietary fiber. *Metabolism*. 1993;42:585-93.

[53] Stark, A. H., Crawford, M. A., Reifen, R. Update on alpha-linolenic acid. *Nutr. Rev.* 2008;66:326-32.
[54] Glaser, C., Lattka, E., Rzehak, P., Steer, C., Koletzko, B. Genetic variation in polyunsaturated fatty acid metabolism and its potential relevance for human development and health. *Matern. Child Nutr.* 2011;7 Suppl. 2:27-40.
[55] Mattar, M., Obeid, O. Fish oil and the management of hypertriglyceridemia. *Nutr. Health.* 2009;20:41-9.
[56] Caslake, M. J., Miles, E. A., Kofler, B. M., Lietz, G., Curtis, P., Armah, C. K., et al. Effect of sex and genotype on cardiovascular biomarker response to fish oils: the FINGEN Study. *Am. J. Clin. Nutr.* 2008;88:618-29.
[57] Maki, K. C., Dicklin, M. R., Davidson, M. H., Doyle, R. T., Ballantyne, C. M. Baseline lipoprotein lipids and low-density lipoprotein cholesterol response to prescription omega-3 acid ethyl ester added to Simvastatin therapy. *Am. J. Cardiol.* 2010;105:1409-12.
[58] Olano-Martin, E., Anil, E., Caslake, M. J., Packard, C. J., Bedford, D., Stewart, G., et al. Contribution of apolipoprotein E genotype and docosahexaenoic acid to the LDL-cholesterol response to fish oil. *Atherosclerosis.* 2010;209:104-10.
[59] Minihane, A. M., Khan, S., Leigh-Firbank, E. C., Talmud, P., Wright, J. W., Murphy, M. C., et al. ApoE polymorphism and fish oil supplementation in subjects with an atherogenic lipoprotein phenotype. *Arterioscler. Thromb. Vasc. Biol.* 2000;20:1990-7.
[60] Lindi, V., Schwab, U., Louheranta, A., Laakso, M., Vessby, B., Hermansen, K., et al. Impact of the Pro12Ala polymorphism of the PPAR-gamma2 gene on serum triacylglycerol response to n-3 fatty acid supplementation. *Mol. Genet. Metab.* 2003;79:52-60.
[61] Ylonen, S. K., Salminen, I., Lyssenko, V., Virtanen, S. M., Groop, L., Aro, A., et al. The Pro12Ala polymorphism of the PPAR-gamma2 gene affects associations of fish intake and marine n-3 fatty acids with glucose metabolism. *Eur. J. Clin. Nutr.* 2008;62:1432-9.
[62] Auinger, A., Helwig, U., Rubin, D., Herrmann, J., Jahreis, G., Pfeuffer, M., et al. Human intestinal fatty acid binding protein 2 expression is associated with fat intake and polymorphisms. *J. Nutr.* 2010;140:1411-7.
[63] Pishva, H., Mahboob, S. A., Mehdipour, P., Eshraghian, M. R., Mohammadi-Asl, J., Hosseini, S., et al. Fatty acid-binding protein-2 genotype influences lipid and lipoprotein response to eicosapentaenoic acid supplementation in hypertriglyceridemic subjects. *Nutrition.* 2010 Jan 15.
[64] Khedara, A., Kawai, Y., Kayashita, J., Kato, N. Feeding rats the nitric oxide synthase inhibitor, L-N(omega)nitroarginine, elevates serum triglyceride and cholesterol and lowers hepatic fatty acid oxidation. *J. Nutr.* 1996;126:2563-7.
[65] Piolot, A., Blache, D., Boulet, L., Fortin, L. J., Dubreuil, D., Marcoux, C., et al. Effect of fish oil on LDL oxidation and plasma homocysteine concentrations in health. *J. Lab. Clin. Med.* 2003;141:41-9.

[66] Ferguson, J. F., Phillips, C. M., McMonagle, J., Perez-Martinez, P., Shaw, D. I., Lovegrove, J. A., et al. NOS3 gene polymorphisms are associated with risk markers of cardiovascular disease, and interact with omega-3 polyunsaturated fatty acids. *Atherosclerosis*. 2010;211:539-44.

[67] Katan, M. B., Grundy, S. M., Jones, P., Law, M., Miettinen, T., Paoletti, R. Efficacy and safety of plant stanols and sterols in the management of blood cholesterol levels. *Mayo. Clin. Proc.* 2003;78:965-78.

[68] Calpe-Berdiel, L., Escola-Gil, J. C., Blanco-Vaca, F. New insights into the molecular actions of plant sterols and stanols in cholesterol metabolism. *Atherosclerosis*. 2008.

[69] Thompson, G. R., Grundy, S. M. History and development of plant sterol and stanol esters for cholesterol-lowering purposes. *Am. J. Cardiol.* 2005;96:3D-9D.

[70] Rideout, T. C., Harding, S. V., Mackay, D., Abumweis, S. S., Jones, P. J. High basal fractional cholesterol synthesis is associated with nonresponse of plasma LDL cholesterol to plant sterol therapy. *Am. J. Clin. Nutr*. 2010;92:41-6.

[71] Hallikainen, M. A., Sarkkinen, E. S., Gylling, H., Erkkila, A. T., Uusitupa, M. I. Comparison of the effects of plant sterol ester and plant stanol ester-enriched margarines in lowering serum cholesterol concentrations in hypercholesterolaemic subjects on a low-fat diet. *Eur. J. Clin. Nutr*. 2000;54:715-25.

[72] Vanhanen, H. T., Blomqvist, S., Ehnholm, C., Hyvonen, M., Jauhiainen, M., Torstila, I., et al. Serum cholesterol, cholesterol precursors, and plant sterols in hypercholesterolemic subjects with different apoE phenotypes during dietary sitostanol ester treatment. *J. Lipid Res*. 1993;34:1535-44.

[73] Plat, J., Mensink, R. P. Relationship of genetic variation in genes encoding apolipoprotein A-IV, scavenger receptor BI, HMG-CoA reductase, CETP and apolipoprotein E with cholesterol metabolism and the response to plant stanol ester consumption. *Eur. J. Clin. Invest*. 2002;32:242-50.

[74] Geelen, A., Zock, P. L., de Vries, J. H., Katan, M. B. Apolipoprotein E polymorphism and serum lipid response to plant sterols in humans. *Eur. J. Clin. Invest.* 2002;32:738-42.

[75] Banuls, C., Martinez-Triguero, M. L., Lopez-Ruiz, A., Morillas, C., Jarabo, M. M., Bellod, L., et al. Serum lipid responses to phytosterol-enriched milk in a moderate hypercholesterolemic population is not affected by apolipoprotein E polymorphism or diameter of low-density lipoprotein particles. *Eur. J. Clin. Nutr*. 2011;65:255-61.

[76] Zhao, H. L., Houweling, A. H., Vanstone, C. A., Jew, S., Trautwein, E. A., Duchateau, G. S., et al. Genetic variation in ABC G5/G8 and NPC1L1 impact cholesterol response to plant sterols in hypercholesterolemic men. *Lipids*. 2008;43:1155-64.

[77] Rudkowska, I., Abumweis, S. S., Nicolle, C., Jones, P. J. Association between non-responsiveness to plant sterol intervention and polymorphisms in cholesterol metabolism genes: a case-control study. *Appl. Physiol. Nutr. Metab*. 2008;33:728-34.

[78] Plat, J., Bragt, M. C., Mensink, R. P. Common sequence variations in ABCG8 are related to plant sterol metabolism in healthy volunteers. *J. Lipid Res*. 2005;46:68-75.

[79] Gylling, H., Hallikainen, M., Raitakari, O. T., Laakso, M., Vartiainen, E., Salo, P., et al. Long-term consumption of plant stanol and sterol esters, vascular function and genetic regulation. *Br. J. Nutr.* 2009;101(11):1688-95.

[80] Rabot, S., Rafter, J., Rijkers, G. T., Watzl, B., Antoine, J. M. Guidance for substantiating the evidence for beneficial effects of probiotics: impact of probiotics on digestive system metabolism. *J. Nutr.* 2010;140:677S-89S.

[81] Ma, K. Y., Yang, N., Jiao, R., Peng, C., Guan, L., Huang, Y., et al. Dietary calcium decreases plasma cholesterol by down-regulation of intestinal Niemann-Pick C1 like 1 and microsomal triacylglycerol transport protein and up-regulation of CYP7A1 and ABCG 5/8 in hamsters. *Mol. Nutr. Food Res.* 2011;55:247-58.

[82] Butler, J. A., Hagen, T. M., Moreau, R. Lipoic acid improves hypertriglyceridemia by stimulating triacylglycerol clearance and downregulating liver triacylglycerol secretion. *Arch. Biochem. Biophys.* 2009;485:63-71.

[83] Khoo, Y. S., Aziz, Z. Garlic supplementation and serum cholesterol: a meta-analysis. *J. Clin. Pharm. Ther.* 2009;34:133-45.

[84] Marinangeli, C. P., Jones, P. J., Kassis, A. N., Eskin, M. N. Policosanols as nutraceuticals: fact or fiction. *Crit. Rev. Food Sci. Nutr.* 2010;50:259-67.

In: Nutrigenomics
Editor: M. T. Ravi Subbiah

ISBN: 978-1-62417-051-5

Chapter VI

Nutraceuticals to Prevent Thrombogenesis and Its Application to Women's Health

M. T. Ravi Subbiah,[1] S. P. Bydlowski[2] and L. Maselli[3]
[1]Department of Internal Medicine, University of Cincinnati Medical Center, Cincinnati, Ohio, US
[2]Department of Hematology, University of Sao Paulo School of Medicine, Sao Paulo, Brazil
[3]Molecular Genetics and Biotechnology Laboratory, Research Division, Pro Sanguo Hemocentro de Sao Paulo Foundation, Sao Paulo, Brazil

Abstract

Thrombosis plays an important role in diseases affecting the arterial and venous vascular systems and substantially contributes to morbidity and mortality in Western countries. The etiology of both arterial and venous thrombosis is complex and involves environmental and genetic factors. Atherothrombosis represents an advanced stage of atherosclerosis in which plaque rupture or erosion trigger increased clotting activity, leading to the formation of an arterial thrombus. Platelets play a pivotal role in the development of atherosclerosis. This important contribution of platelets is strengthened by the observed benefits of antiplatelet agents like aspirin and clopidogrel in primary and secondary therapy of atherothrombotic disorders. Advances in human genome project have enabled scientific investigators to identify a number of mutations in genes that increase the risk of thrombosis. These genetic tests are routinely carried out in the laboratory to identify individuals that may have increased risk for thrombosis and they come under the 'thrombophilic panel' of tests. There are some issues unique to women' health that deserves increased attention for the risk of thrombosis. Oral contraceptive use in women is associated with increased risk of deep vein thrombosis and stroke and risk is even greater for those women with prothrombin and factor V Leiden mutations. Similarly, hormone replacement therapy (HRT) in postmenopausal has also indicated the negative effects on the risk of stroke and venous thrombosis. Women's Health Initiative (WHI) trial using estrogen plus progestin reported a doubling of the risk of stroke and venous thrombosis. These observations provide opportunities to develop nutraceuticals to reduce the activity of platelets and the risk of thrombosis. Antioxidants, procyanidins from guarana (paulinnia cupana), polyphenols from grape and other berries, green tea and garlic phytochemicals are being explored to as nutraceuticals to combat thrombosis. Such nutraceuticals may play a preventive role as dietary supplements, especially in women taking oral contraceptives in reproductive years or hormone replacement therapy in menopause.

Keywords: venous thrombosis, atherothrombosis, thrombin, platelet aggregation, genetic tests, thrombogenic panel, methylene tetrahydofolate reductase, protein C, protein S, Factor V Leiden mutation, prothrombin mutation, hyperhomocysteinemia, women's health, hormone replacement therapy, plasminogen activator-inhibitor-1, coronary heart disease, anti-platelet aggregation agents, guarana, catechin, epicatechin, caffeine, antioxidants, allicin, polyphenols, echistatin, folic acid, thromboxane A2, thromboxane B2, ω3 fatty acids

Introduction

Thrombosis represents a major complication for atherosclerosis and the resulting disease process called atherothrombosis is the leading cause of coronary artery disease, stroke and their debilitating health effects [1, 2]. Platelets play an important role in thrombus formation through their ability to: a) activate, aggregate and form clots in response to endothelial injury; b) interact with injured endothelium to release a host of autocrine and paracrine signaling molecules that stimulate smooth muscle cell growth, recruit monocytes and increase inflammation; and thereby contribute towards foam cell formation and accelerate atherogenesis [3-5]. Thrombosis is also a major problem subjects with defects in the coagulation factors and genetic mutations [9-10]. Anticoagulation therapy with agents that target thrombin and decrease fibrin production are routinely used to combat this problem [11, 12]. The completion of the Human genome project [13] has enabled scientific investigators to identify a number of mutations in genes that increase the risk of thrombosis. These genetic tests are routinely carried out in the laboratory to identify individuals that may have increased risk for thrombosis and they come under the 'thrombophilic panel' of tests. In this chapter we discuss a) underlying causes of thrombosis; b) the mechanisms involved; c) role of genetic mutations and their particular significance in women's health and d) how nutraceuticals can potentially benefit women in decreasing platelet aggregation and thrombosis risk.

Thrombosis is the underlying cause of several common diseases involving the arterial and venous vascular systems, and substantially contributes to morbidity and mortality in Western countries [14]. The etiology of both arterial and venous thrombosis is complex and involves environmental and genetic factors.

Atherothrombosis represents an advanced stage of atherosclerosis in which plaque rupture or erosion trigger increased clotting activity, leading to the formation of an arterial thrombus [14, 15]. Platelets are essential for primary hemostasis, repair of damaged endothelium, and play a pivotal role in the development of atherosclerosis [16]. In contrast, the combination of stasis and hypercoagulability are much more crucial than endothelial damage and activation for the occurrence of venous thrombosis. Venous thrombi are mainly constituted by fibrin and red blood cell, and less by platelets. Classical concepts proposed by Virchow in 1857 continued to hold good even today, with a triad of causes for venous thrombosis, postulating that stasis, changes in the vessel wall or changes in the blood could lead to thrombosis [17]. Incident venous thromboembolism (VTE) is triggered by a confluence of situational (e.g., trauma, surgery, cancer, infection, inflammation) and genetic factors. The inflammatory response triggered by infection results in a net pro-coagulant effect by concurrent activation of coagulation, driven by tissue factor mediated thrombin generation, and impairment of anticoagulant mechanisms, most notably the protein C system [18]. Inflammation elicits coagulation primarily by activating the tissue factor pathway [19]. Tissue factor binds and activates clotting factor VII, which via factor X results in the generation of thrombin and fibrin.

Normally, the coagulation process is under control of several inhibitors that limit clot formation near the damaged vessel wall, thus avoiding thrombus propagation. This delicate balance can be interrupted whenever an increased pro-coagulant activity of one of the coagulation factors or a decreased activity of one of the naturally occurring inhibitors takes place, leading to thrombus formation.

Venous thrombosis has an age-dependent incidence of one to three individuals per 1000 per year [20]. Unlike atherothrombosis, the role of monogenic conditions leading to venous thromboembolism are well established.

Pro-coagulant activity is regulated by three important anticoagulant pathways: antithrombin, tissue factor pathway and the protein C system. Anti-thrombin, protein C and protein S deficiencies are rare but strong risk factors for venous thrombosis. They have little or no effect on arterial thrombosis.

While the deficiencies of the natural anticoagulants (anti-thrombin, protein C, and its cofactor protein S) may be viewed as strong risk factors (risk increase from 5- to 10-fold), and factor V Leiden and prothrombin 20210A as moderate risk factors (risk increase 2- to 5-fold), several genetic factors, such as the variants in methyltetrahydrofolate reductase (MTHFR), are known to have a weak effect on thrombosis (relative risks between 1.0 and 1.5) [21]. Thrombophilia can be defined as a clinical condition of high propensity, or increased risk, to develop thrombosis, due to predisposing risk factors that may be genetically determined, acquired or both [22].

Although thrombosis may occur in either or both venous and arterial vessels, thrombophilia is usually considered within the context of venous thrombosis. Although arterial thromboses have been reported in protein C, protein S, antithrombin deficient patients and in those with factor V Leiden, components of hereditary thrombophilia [23], the role of thrombophilia in arterial thrombosis remains a controversial issue. Acquired predisposition to thrombophilia includes autoimmune diseases such as antiphospholipid syndrome and systemic lupus erythromatosis.

Other risk factors include prolonged immobility, increasing age, surgery, trauma, cancer, obesity, hyperhomocysteinemia, poor nutrition, pregnancy, puerperium, use of oral contraceptives and hormone replacement therapy.

Hereditary predisposition to thrombosis includes deficiencies of anticoagulant molecules such as protein C, protein S, and antithrombin. In addition, genetic polymorphisms or mutations such as prothrombin mutation G20210A (FII G20210A), Factor V Leiden, that give rise to the condition known as activated protein C resistance, and mutations of fibrinogen could also lead to thrombosis.

In fact, the two most common genetic risk factors for venous thrombosis are the G1691A mutation in the factor V gene (factor V Leiden) and the G20210A mutation in the prothrombin gene.

Factor V Leiden

Factor V (FV) is a protein of the coagulation cascade. The gene encoding human factor V is composed of approximately 80 Kb and is localized on the long arm of chromosome 1. The gene comprises 25 exons and 24 introns, which transcribe into 6.8 Kb of mRNA, which encodes a chain of 2224 amino acid residues with 28 signal peptide amino acids that are removed.

Molecules of factor V are in plasma as single chain glycoprotein composed of A1-A2-B-A3-C1-C2 domains, an identical organization structure as factor VIII. Inactivated FV has a low procoagulation activity until activated by thrombin or activated factor X (FXa) by cleaving in the sites Arg709, Arg1018 and Arg1545. This activated factor V (FVa) molecule is a heterodimer. FVa catalyses FXa mediated conversion of prothrombin to thrombin in the prothrombinase complex. The mechanism by which FVa acts on FXa is not completely understood. The co-factor activity of FVa is balanced by activated protein C (APC), which cleaves FVa at the Arg306, Arg506 and Arg679 sites of the heavy chain. Aside from its procoagulation function, FVa also possesses an anticoagulation function, which is expressed in the case of APC activation by stimulating proteolytic cleavage of activated factor VIII (FVIIIa). In the inactive form FV acts as APC co-factor in the regulatory activity of FV. In inherited and acquired defects, this double role allows FV to influence the manifestation of these disorders into hemorrhagic or thrombotic forms.

FV Leiden refers to a single base change in the FV gene (factor V G1691A) that encodes an amino acid substitution (R506Q). This substitution of an arginine by glutamine at the position 506 eliminates one of three activated protein C cleavage sites in factor V. Therefore, mutant factor V is resistant to inactivation by activated protein C. As a result, FV is inactivated at a lower rate and persists longer in the circulation, leading to more thrombin generation, which explains the procoagulation states in carriers of the Leiden mutation. Factor V mutation was first described by Bertina and his colleagues in 1994 at the University of Leiden [24], based on the discovery of resistance to activated protein C, which was first described in 1993. Among the Caucasian population, factor V Leiden is the most common genetic defect causing thrombosis. The frequency of this mutation in Caucasians ranges between 2 and 15%. A single FV Leiden allele is present in about 5%, 2.2%, and 1.2% of individuals in Caucasian, Hispanic, and African American populations, respectively; overall, about 1 in 5000 individuals is homozygous for the mutation [25, 26].

Factor V Leiden is present in 20% of patients with venous thrombosis and in over half of probands in selected families with thrombophilia. This means that it represents the most common genetic abnormality in patients with thrombosis. The heterozygous form of factor V Leiden increases the risk of thrombosis 3-8 times, while the homozygous form presents a risk which may be up to 80 times greater. Despite their hypercoagulative state, FV Leiden carriers have only a 10% lifetime risk of thrombosis, and the majority of them never develop clinical symptoms. Therefore, the FV Leiden mutation can be considered a mild risk factor for venous thrombosis, which only becomes symptomatic in combination with additional genetic and/or acquired risk factors. Several thrombophilic defects, including protein C deficiency, protein S deficiency, and the prothrombin G20210A mutation, have shown to co-aggregate with FV Leiden in thrombophilia families. Among the acquired factors, the use of oral contraceptives strongly interacts with FV Leiden, increasing the risk of Young female carriers from 8-fold in nonusers to >30-fold in users. In addition to FV Leiden, several other mutations and polymorphisms in the FV gene have been shown to cause plasma activated protein C resistance (APC) in around 5% of individuals. The most important are mutations at the Arg 306 cleavage site (FV Cambridge (G1091C, Arg 306 Thr), FV Hong Kong (A1090G, Arg 306 Gly)), FV Liverpool (T1250C, Ile 559 Thr), FV Glu 666 Asp and FV R2. However, the nature of the genetic variants responsible for APC resistance in the absence of FV Leiden remains largely unknown.

Prothrombin 20210A

The prothrombin (Factor II) mutation is the second most common inherited risk factor for thrombosis. The G20210A mutation in the gene is associated with an elevation of prothrombin levels to about 30% above normal in heterozygotes and to 70% above normal in homozygotes. This mutation is present in about 2% of the Caucasian population, with many variations, such as a higher prevalence of mutation in southern than in northern Europe. The prevalence of the mutation in patients with a first episode of venous thromboembolism is around 6% or more. Six to 12% of individuals who are heterozygous for FV Leiden and have venous thromboembolism also have the prothrombin G20210A mutation (double heterozygotes) [25, 27]. The G20210A mutation in the prothrombin gene is a G-to-A transition at nucleotide position 20210 in the 3'-untranslated region of the coagulation prothrombin gene. The mechanism of action of this mutation is explained by its ability to cause an increase in the level of plasma prothrombin, heightened thrombin formation and increase the risk of developing venous thrombosis [28].

Hyperhomocysteinemia

Homocysteine is an aminoacid with a thiol (-SH) group formed as an intermediate product in the methionine cycle, the function of which is to generate one-carbon methyl groups for transmethylation reactions essential to all forms of life [29].

Hyperhomocysteinemia is a mild risk factor for thrombosis attributable to an impairment of the metabolic pathway that transform the amino acid methionine into cysteine. Slighty increased levels of homocysteinemia are associated with increased risk of thrombosis [30]. It is considered to increase the risk of venous thromboembolism about 2.5-fold. Overall, there appears to be a weak positive association between arterial thrombosis and hyperhomocysteinemia. Whether hyperhomocysteinemia is a marker or a cause of thrombosis has long been debated. Although hyperhomocysteinemia can be treated easily with vitamin supplementation (folate, vitamin B_6, and vitamin B_{12}) [31], reduction of total homocysteine was not shown to reduce the risk of recurrent venous thromboembolism [32]. Therefore some experts doubt the value of total homocysteine as a component of thrombophilia testing.

Hyperhomocysteinemia may develop due to genetic or acquired dispositions. Acquired disposition primarily include low vitamin intake (B_{12}, B_6, folic acid).

Genetic factors include the very rare cystathionine-β-synthase deficit, whose homozygous form represents classical hyperhomocysteinemia and leads to very high levels of homocysteinemia [33]. Severe homocystinuria is an inherited disease due to homozygous or compound heterozygous mutations in cystathionine-β-synthase and other enzymes regulating methionine metabolism. Over 130 mutations of the cystathionine-β-synthase gene have been reported [29]

The very common variant of the methyltetrahydrofolate reductase (MTHFR) gene, a C to T transition at nucleotide 677, leads to the thermo-labile enzyme variant with impaired activity and a slightly increased homocysteinemia level [34]. Currently, direct association between the MTHFR 677TT variant, increased homocysteinemia level and risk of developing thrombosis is not distinctly established. It has been shown that hyperhomocysteine is present only in subjects with the 677TT genotype and concomitant folate deficiency, which often occurs in the eldery [35]. Mutations in the 5, 10-methylene tetrahydrofolate reductase (MTHFR) and methionine synthase (MS) genes are less frequent but can also cause homocystinuria.

Antithrombin Deficiency

Deficiency of antithrombin (AT) is associated with an increased risk of venous thrombosis and pulmonary embolism. Until 1965, it had been believed that antithrombin III (ATIII) was the inhibitor in plasma that slowly degraded thrombin, whereas ATII was the fast-acting AT that was generated when heparin was added to plasma. Now is established that both activities are different properties of the same 432-amino acid mature protein. We now refer to AT deficiency (rather than ATIII deficiency), and deficiency of AT is the most clinically severe deficiency of a natural anticoagulant [36].

Antithrombin directly inhibits several activated coagulation factors, particularly thrombin and activated factor X. Its deficiency results in significant reduced inhibition of thrombin and activated factor X and an increased tendency to clot formation, particularly in the venous system, where the coagulation pathway (as distinct from platelets) plays a more important role in thrombus formation.

The 13.57 Kb gene encoding AT, SERPINC1, is located on the long arm of chromosome 1 (q23-q25.1) and contains seven exons and six introns. Several polymorphisms have been identified including an insersion/deletion dimorphism characterized by the presence of a 76-bp fragment in the 5' untranslated region. Over 200 distinct mutations causing AT deficiency have been described [36, 37] . AT deficiency is a heterogeneous disorder that usually shows an autosomal dominant pattern of inheritance. Most patients are heterozygous for the defect. Complete quantitative deficiency is thought to be incompatible with life and results in embryonic lethality in mice [38].

Although very rare, the thrombotic risk for patients with antithrombin deficiency was shown to be increased by 5-fold. Despite this high risk, antithrombin deficiency represents only 1% of all thrombotic conditions. AT deficiency is a significantly more serious risk factor for developing thrombosis than protein C and S deficiency.

Antithrombin deficits are divided into two basic groups: acquired and inherited. Acquired deficiency causes include decreased synthesis of antithrombin as in liver cirrhosis, increased losses of antithrombin as in gastrointestinal diseases or nephrotic syndrome, and increased antithrombin consumption as in disseminated intravascular coagulation, sepsis, burns or postoperative conditions. Acquired deficiency may be also present under physiological conditions (newborns, late stages of pregnancy). Inherited antithrombin deficiencies are characterized by a decreased level of antithrombin antigen, activity or both, caused by an inherited defect in the synthesis of antithrombin or in the synthesis of mutant antithrombin molecules. The disorders are autosomal dominantly inherited.

The classification of heritable AT deficiency defines type I and type II deficiency states. Type I deficiency represents 80-90% of all antithrombin defects, where a parallel decrease in AT activity and AT antigen in plasma is observed. The deficiency is typically caused by reduced synthesis and secretion of qualitatively normal AT into the blood. The levels of antithrombin decreased to 50-70% of normal values. It is usually caused by nonsense mutations or small insertions and deletions within SERPINC 1 which lead to frame shift and result in a failure expression of AT. Missense mutations that result in quantitative deficiency have also been described.

Type II deficiency is characterized by a misbalance between antigen concentration and antithrombin functional activity, the antigen level is normal, but the molecules have a functional defect. It is usually caused by missense mutations affecting residues that are involved in AT function. Such types of AT defects include decreased affinity to heparin (antithrombin Budapest, antithrombin Milano, antithrombin Tokyo), decreased capability to inactivate serine proteases (antithrombin Northwick Park, antithrombin Glasgow), and defective capacity to bind to heparin (antithrombin Tours, antithrombin Rouen).

Protein C and Protein S

Coagulation FV and FVIII are homologous proteins composed of multiple domains (A1-A2-B-A3-C1-C2). Following limited proteolysis by FXa or thrombin, they lose the B domain and are converted to their active forms. FVa and FVIIIa act as the essential nonenzymatic cofactors of the serum-proteases FXa and FIXa in the activation of prothrombin and FX, respectively. As FVa and FVIIIa controls two key steps of the coagulation cascade, their activity needs to be tightly regulated [39]. This is done by the protein C (PC) anticoagulant system, in which activated protein C (APC), together with its cofactors protein S and FV, catalyses the proteolytic inactivation of FVa and FVIIIa.

The protein C anticoagulant pathway is a major anticoagulant system. It plays a crucial role in the regulation of fibrin formation via the proteolytic inactivation of procoagulant factors Va and VIIIa, the essential cofactors of the prothrombinase and tenase complexes, respectively [40]. PC is a vitamin K-dependent plasma glycoprotein that circulates in plasma as an inactive zymogen, while is converted to the active serine protease (activated PC) by limited proteolysis of a single peptide bond [41]. The gene for human protein C is located on chromosome 2 (q13-q14). It comprises nine exons which code for a 419-amino acid mature protein highly homologous to prothrombin and factors VII, IX, and X. The activation takes place on the surface of endothelial cells by the thrombin-thrombomodulin complex [42]. PC is activated by thrombin bound to thrombomodulin. Thrombin cleaves a single peptide bond (Arg160) in the heavy chain of protein C, releasing an activation peptide of 12 amino acids. APC subsequently inactivates FVa and FVIIIa by limited proteolysis at specific sites. APC cleaves FVa at Arg 306, Arg 506 and Arg 679 residues in the heavy chain. Although Arg 506 is the kinetic favoured cleavage site, cleavage at Arg 306 is required for complete FVa inactivation. Both inactivation reactions are stimulated by its main cofactor, protein S [22]. Moreover, FV acts as an additional APC-cofactor in the inactivation of FVIIIa.

This reaction is greatly stimulated by the endothelial surface protein C receptor (EPCR) which binds protein C and brings it in close proximity with the thrombin/ thrombomodulin complex [43]. This binding enhances the rate of PC activation by the thrombin-thrombomodulin complex by 5- to 20-fold. The EPCR belongs to the major histo-compatibility complex (MHC) class I/CD1 family of membrane proteins. Once activated, APC may dissociate from EPCR, bind to its cofactor protein S, and exhibit its anticoagulant function.

FVa inactivation is enhanced by the APC-cofactor protein S [44]. Protein S is an anticoagulation protein that acts as a non-enzymatic co-factor of activated protein C during inactivation of factors Va and VIIIa. Protein S is a 75-kDa vitamin K-dependent glycoprotein. The gene for protein S spans ~80Kb on chromosome 3 (q11.2) and contains 15 exons. Mature protein S is a 635-amino acid protein. Free protein S enhances APC-catalyzed cleavage of FVa at Arg306 by 20-fold, whereas it only has a minor effect on cleavage at Arg506. By replacing Arg506 with a Glu, the FV Leiden mutation abolishes the preferential cleavage site recognized by APC on FVa. Therefore, protein S is particularly important for efficient inactivation of FVa Leiden. Differently, protein S only marginally accelerates cleavage of Arg 506. The inherited deficiency of one of these inhibitors leads to a critical reduction in the natural anticoagulant system and enhances thrombin generation, increasing susceptibility to venous thrombosis.

Activated protein C (APC) resistance is an *in vitro* phenomenon characterized by a slight anticoagulation response to APC in plasma that leads to inadequate regulation of thrombin production. As such, APC resistance is associated with an increased risk of developing thrombosis. APC resistance is associated with FV Leiden mutation in up to 90% of cases.The clinical relevance of the PC pathway is evident from reports showing a clear association between deficiencies of PC and protein S or reduced APC levels with thrombosis. In fact, the deficiencies of PC and protein S or the FV Leiden mutation are present in more than 50% of patients with inherited thrombophilia [24]. Normal APC generation depends on the presence assembly of thrombin and PC to their respective receptor, TM and EPCR, on the surface of endothelial cells. Any change in the efficiency of this coupling may cause altered APC generation and a modification in the risk of thrombosis.

Congenital deficiencies of protein C and protein S are rare in the general population. In contrast, functional abnormalities of the protein C pathway are common. The prevalence of protein C deficiency in the population combined with high risk of thrombogenicity results in a 1-3% contribution of protein C deficiency on all thrombophilic conditions. The relative thrombotic risk for protein C deficit is 6.5. The gene causing protein C deficiency is autosomal and dominantly inherited. Heterozygous protein C defects are associated with a 7-fold increase in the risk of thrombosis. Two types of protein C deficiency are described:

quantitative (decrease in antigen) and functional. Functional deficiency may be caused by a number of mutations. To date, more than 160 different mutations have been described. The prevalence of protein S deficiency represents 1-2% of patients with deep venous thrombosis and 6% of families with thrombophilia. More than 70 different mutations have been described in the gene for protein S.

Finally, several mutations and polymorphisms have been reported in the EPCR gene. Whether or not functional polymorphisms in the EPCR gene alter the risk of thrombosis, alone or in combination with other risk factors, has not been undoubtedly demonstrated and requires further investigation.

PAI-1 is a serine protease inhibitor that is the central key component of the fibrinolytic system. During fibrinolysis, plasminogen is converted to plasmin, which in turn dissolves fibrin clots. Two physiologic plasminogen activators (PA) have been described, tissue type (tPA) and urokinase type (u-PA). This conversion is regulated to a significant extent by PAI-1, which inhibits the fibrinolytic system. Because of its major role in the regulation of the fibrinolytic process, PAI-1 overexpression may affect normal fibrin clearance mechanisms, thus leading to pathological fibrin deposition and thrombotic events [45]

The human PAI-1 is synthesized in several tissues including liver and endothelial cells. The human PAI-1 gene is localized at chromosome 7q22. A genetic polymorphism of this gene has been identified in the promoter region, at -675 bp upstream from the mRNA initiation point, where one allele has a sequence of four guanosines (4G) and the other has five (5G) [46].

Both the 4G and 5G alleles have a transcription activator binding site. However, the 5G allele has an additional binding site for a repressor, leading to lower transcription rates and, consequently, less PAI-1 activity. Therefore, the 4G allele has been linked with moderately higher plasma PAI-1 levels [47]. Data regarding the association between the presence of allele 4G and the risk for arterial or venous thrombotic episodes have been controversial. However, it seems that the presence of the 4G allele might significantly increase the thrombotic risk in patients with thrombophilia [48].

Thrombosis and Its Significance in Women's Health

The increase in the risk for venous and arterial thrombosis following oral contraceptive use is well established [49, 50]. This risk was further enhanced venous in women with mutations in prothrombin gene and increased factor VIII [51-53]. Studies by Kemmeren et al. [54] compared second (ethanol estradiol/levonorgestrel) and third (ethinyl estradiol/desogestrel) generation oral contraceptives in women with factor V Leiden mutation and noted higher thrombotic risk in both regimens with desogestrel group showing a more pronounced risk. A European case control study [55] noted that in women who smoke, oral contraceptives increased the risk of stroke as well.

Similarly, hormone replacement therapy (HRT) in postmenopausal women has also indicated the negative effects on the risk of stroke and venous thrombosis [55, 56]. Women's Health Initiative (WHI) trial using estrogen plus progestin reported [56] an increase in risk of all stroke was apparent in this study. Cushman et al. [57] reported that estrogen/progestin doubled the risk of stroke and venous thrombosis using the final data from the WHI study. It is interesting to note that increased thrombosis risk was not evident in esterified estrogen (as opposed to conjugated estrogen) used in WHI study [58]. Similarly, negative effect of estrogen therapy on cardiovascular risk was also reported using data derived from WHI study [59-61].The mechanisms involved in the observed increase in thrombosis and CV events following estrogen therapy is not clear. A recent cross-sectional study [62] of interactions between the thrombophilic Factor V Leiden gene mutation, ERT and CHD in women provides insight into the failure of ERT to reduce CHD in the three major trials [59-61] discussed above. The investigators reported an interaction between ERT-mediated tendency to form blood clots and the Factor V Leiden mutation (present in 6% of American women). They speculated that when ERT-mediated thrombophilia is superimposed on the heritable thrombophilic Factor V Leiden mutation, CHD is promoted and any putative ERT associated reduction in CHD is overshadowed. Recently, ERT interaction with another commonly heritable trait, the prothrombin gene mutation was suggested [63]. In women referred for therapy of hyperlipidemia, the authors concluded that when the prothrombin G20210A is absent ERT is protective against CHD, but in the presence of this mutation ERT may actually increase risk for CHD. Similar findings have been reported by Psaty et al. [64]. These studies suggest that a gene-gender-HRT interaction accounts for increased CHD events in women on HRT, arising from a small sub-set of women (about 12%), heterozygous for either the Factor V Leiden and/or prothrombin gene mutations. This interaction could be particularly strong if they are also homozygous for hypofibrinolytic 4G/4G polymorphism of the plasminogen activator inhibitor (PAI-I) gene as previous studies have demonstrated that 4G/4G genotype of PAI-1 gene is an independent risk factor for CHD [65]. All these studies strongly support the possibility that the increased risk noted in WHI trial may be related with thrombophilic mutations in some study participants. A recent study by Botto et al. [66] investigated to see whether a positive family history of thrombosis is able to identify a genetic predisposition to thrombosis as evaluated by genetic testing for thrombophilic mutations. They noted that both factorV Leiden and prothrombin mutations was significantly higher in asymptomatic women with a positive family history. This indicates that at the least, women with a family history of thrombosis should undergotesting for thrombophilic gene mutations prior to contraceptive or hormone replacement therapy. Furthermore, accumulating evidence for the potential role of obesity and metabolic syndrome as risk factors for venous and arterial thrombosis [67] warrant further monitoring of thrombosis risk factors in obese women. There is also some discussion as to the relationship between the type of estrogen used in hormone replacement therapy and the noted increase in cardiovascular risk. [68] Above mentioned studies indicate young women who take oral contraceptives or postmenopausal women on estrogen replacement therapy, especially with a family history of thrombosis and /or the thrombophilic mutations (Factor V Leiden, Prothrombin and PAI-1 mutations) could benefit by preventive dietary and /or drug therapies that could reduce or control thrombosis risk. Aspirin is a commonly used drug that prevents platelet aggregation and reduces thrombotic risk, but many individuals' exhibit side effects including GI bleeding [69]. Other studies [70] have also noted that there is a gender specific difference in response to aspirin by women who display a higher 'aspirin resistance' when compared to men. Therefore, there is considerable potential for development of nutraceuticals that can help in modulate platelet reactivity and decrease thrombosis risk in women. Nutrigenomics has revitalized the field by demonstrating that nutrients can interact with the genome altering gene expression, enabling their use as 'nutraceutical' for a 'specific metabolic problem'. In this context, we will review compounds that come under the category of viable nutraceuticals towards decreasing risk of thrombosis in women. We will initially discuss our work on guarana, an Amazonian herb followed by review of other natural compounds with ant-platelet and anti-thrombogenic potential.

Potential of Guarana Components As Antithrombotic Nutraceuticals

Guarana (Paulinnia cupana (Family: Saponidaceae, Genus Paulinnia) is a woody spine or sprawling shrub native to the Amazon basin was used extensively by Guarani and Tupi Indians in the Amazon region [71]. Other Synonyms in use include guaranastraik, quarana, and Brazilian cocoa. Guarana has been used as a tonic, astringent and against gastrointestinal ailments in the Amazon. The guarana fruit approximates the size of coffee berry with a black seed. The seeds of guarana fruits are crushed and commonly used as a stimulant and for other health benefits. Guarana seeds are one of the richest natural sources of caffeine amounting to 2-5% of the dry weight of its seeds [72]. Furthermore, unlike the immediate high noted with caffeine (derived from coffee), guarana seeds displayed a steady and sustained release of the stimulant (originally called guaranine). This led to some confusion initially whether the stimulatory substance in guarana seeds was actually caffeine. Subsequent studies showed that 'guaranine' was actually caffeine and no significant difference in its absorption rate (compared to caffeine) was evident [73, 74]. In addition to caffeine, guarana seeds contain smaller amounts of other methyl xanthines, theophylline and theobromine [72]. Other compounds in guarana seeds include carbohydrates, resins and tannins, methyl benzenes, cyclic sesqui-terpene hydrocarbons, fatty acids and plant sterols [75, 76]. Guarana cultivation continues to thrive in the Amazon region under the FUNAI (Indian National Foundation) and it maintains a healthy market for use as a high stimulant food additive both in Europe and US. Although guarana has been in the nutritional supplement market for 40 years, limited knowledge exists with regards to its pharmacology and metabolic effects. While the indigenous people of the Amazon basin believed that guarana was a source of energy and helps maintain good health, further research has shown that guarana or guarana/ginseng combination significantly improved mood and cognitive performance, increased metabolic activity and weight loss in humans [77-79]. Animal and in vitro studies on guarana seed extracts have generally provided support for its role in chemo-protective effects and ability to protect lipid peroxidation.

Anti-Platelet Aggregation Effect of Guarana Seeds

Nearly two decades ago our laboratory carried out studies [80] which showed that extracts of guarana seeds possessed strong inhibitory properties against ADP and arachidonic acid induced platelet aggregation. Other studies by Bydlowski et al. [81] that showed that the anti-platelet aggregatory property of guarana seeds was due to its ability to inhibit thromboxane synthesis. Only recently further investigations were done studies to determine the nature of compounds in guarana seeds responsible for this inhibitory activity [82].

Differential extraction of guarana seeds with different solvent combinations were helpful in suggesting the possible active compounds responsible for platelet inhibition activity (Figure 1). These studies showed that extraction of guarana seeds with butanol did not preserve its platelet aggregation inhibitory activity, but chloroform: methanol (2:1, v/v) extraction to remove lipids) had no effect, indicating butanol solubility of active components.

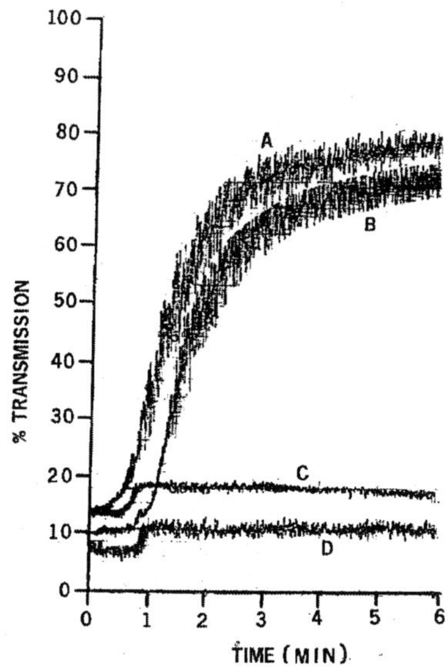

Reproduced with permission from Int. J. Vitam Nutr. Res. 78:96-101, 2009.

Figure 1. Effect of guarana seed fractions on the response of human platelets to ADP. Legends. Response of human platelet rich plasma to ADP (5 ug) and the effect of guarana fractions. A = control; B = guarana extract after butanol extraction; C = lipid-free guarana extract; D = guarana extract free of methyl xanthines. Experimental conditions as described under methods.

Previous studies had shown [80] that platelet aggregation inhibitory activity still persisted even after removal of methyl xanthine fractions (caffeine) by thin-layer chromatography of guarana seed extracts. Upon preparatory thin layer chromatography of total guarana seed extract, about 8 major bands were and a large area containing methyl xanthine were noted.

While some platelet aggregation inhibitory activity was evident in Bands 2 and 4, bulk of the activity was associated with Band 8 (just below the caffeine-rich fraction). HPLC/MS analysis of the purified Band 8 noted significant amounts of catechin, epicatechin along with caffeine and materials eluting at 25-30 min retention time that ionized extensively (Figure 2).

These compounds (with a retention time of 25-30 min) correspond to dimers of catechins (m/e 575). Purified Band 8 even after repeated purification still contained caffeine along with catechin and epicatechin. Further subfractionation of Band 8 by thin layer chromatography resulted into three bands (Bands A, B and C) While Bands A and B contained catechin, epicatechin and their dimers in different proportions, Band C predominantly contained caffeine. Human platelet thromboxane synthesis in vitro were significantly inhibited by Bands A and B, while Band C had a weak effect (Table 1). In general, the available studies indicate no significant toxic effects of guarana seed extracts when administered alone to animals and humans [76, 83]. However, strong adverse reactions have been reported when guarana seeds were co-administered with other selected herbs particularly ephedra-like alkaloids with guarana seeds prolonged the plasma half-life of ephedrine [84]. In some subjects, guarana seed/ephedrine combination increased systolic pressure and mean heart rate after 90 minutes [85].

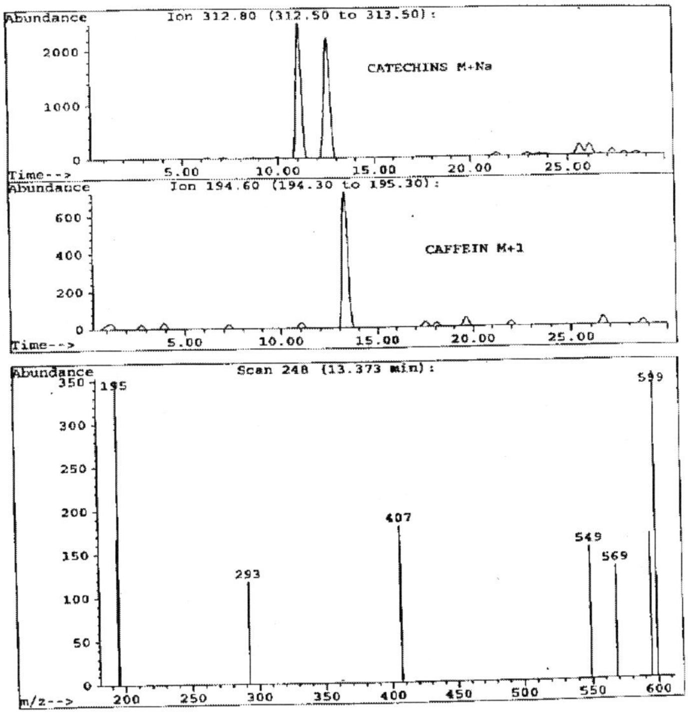

Reproduced with permission from Int. J. Vitam. Nutr. Res. 78:96-101, 2009.

Figure 2. HPLC/MS analysis of Band B isolated from guarana seed extracts Legends. HPLC/MS analysis of Band B using positive electrospray conditions. Note major ions representing molecular weights 312 (289 + 23 for Na) of catechin (first peak) and epicatechin (peak 2) in the top panel. In the middle panel, caffeine with a mass of 195 (mass + 1) is identified. The lower panel shows the presence of other ions notably at 549-599 range perhaps corresponding to dimers of catechins.

Table 1. Effect of sub-fractions of Band 8 isolated from guarana extracts on prostaglandin E2 and thromboxane B2 synthesis in washed human platelets*

Fraction	Thromboxane B2 (dpm/mg protein)	PGE2 (dpm/mg protein)
Control	13534 ± 474	1645 ± 498
Guarana fractions		
Band A	8201 ± 298**	1052 ± 331
Band B	2989 ± 534**	607 ± 82
Band C	9495 ± 738**	1894 ± 462

Legends to Table 1: *Human platelets were pre-incubated with guarana fractions for 2 min at 37 °C in a prior to the addition of 0.1 uC of ^{14}C-arachidonic acid. After a 10 min further incubation, the. prostaglandins were extracted after the addition of authentic standards of PGE2, and thromboxane B2 as carriers, separated by TLC and radioactivity was determined by scintillation counting.

**All data is presented as Mean ± SEM. Significantly different ($p < 0.05$) form control as estimated by one way ANOVA with Benferroni's multiple comparison post test. (Reproduced with permission from Int J Vitam Nutr Res 78:96-101, 2009).

Neutraceuticals Preventing Platelet Aggregation and Thrombosis

Antioxidants

There is strong evidence that reactive active species (ROS) play an important role in platelet activation, aggregation and thrombotic process [86, 87]. ROS has been detected in activated platelets [88] and these compounds either directly or through their target products like fatty acid hydroxyl peroxides have been shown to modify platelet responses [89]. There is evidence for the role of NADPH oxidase in superoxide production and platelet recruitment [90]. Based on this, use of antioxidants like vitamin E to prevent platelet aggregation and thrombosis has been proposed [90]. While there was a suppression of platelet aggregation with vitamin E, this effect appears to be independent of its antioxidant effect [91]. In clinical studies vitamin E was shown to decrease venous thromboembolism in women [92]. Antioxidants could also work by protecting the inactivation of nitric oxide, a vasodilator secreted by the endothelium [93, 94]. In this context, it is interesting to note that selenium containing glutathione peroxidase-3 (antioxidant defense enzyme) also protects nitric oxide inactivation by lipid peroxides [95]. This also supported by studies showing that selenium supplementation can decrease synthesis of pro-aggregatory compound thromboxane A2 and cause a reciprocal increase in the levels of prostacyclin, a potent inhibitor of platelet aggregation [96]. Results of the benefits of vitamin E in some prospective clinical studies in terms of their effects on coronary spasm and other events have been discussed in a review by Chakrabarti and Freedman [97].

Polyphenols

Polyphenol compounds especially belonging to the group called flavonoids (primarily derived from grape products like wine, juice and seeds) have shown strong promise to serve as nutraceuticals capable of inhibiting platelet aggregation and thrombosis [98, 99]. Some of the potential mechanisms involved appear to be antioxidant capacity [100] and protecting nitric oxide deactivation [99]. Furthermore, it is of interest to note [101] that red wine is more efficient in modulating platelet aggregation than white wine. The major compounds among polyphenols responsible for effects on platelet aggregation appear to be quercetin, epicatechins, and resvertrol [97, 98]. Pignatelli et al. [102] suggested that some of these compounds could be operating by reducing platelet recruitment through the inhibition of PKC-dependent NADPH oxidase activation and/or nitric oxide mediated platelet glycoprotein GpIIb/IIIa down regulation offers interesting possibilities to screen for new and more potent compounds affecting these sites

Some components of tea like epigallocatechin-3 gallate has been shown to reduce platelet aggregation induced by thrombin, ADP and collagen [103] and also inhibit the binding of fibrinogen to the platelet surface GpIIb/IIIa complex by attenuating cytoplasmic calcium concentration [104]. Similarly, phytoestrogen genistein found in soy products can also successfully inhibit platelet aggregation and thrombosis in addition to its anti-angiogenic property [105,106].

Other Componds

Echistatin, a compound isolated from snake venom also has been shown to inhibit platelet aggregation through its property to serve as a competitive inhibitor of platelet αIIbβ3 integrin binding to fibrinogen [107]. Other compound of interest is folic acid, which is suggested to be effective in decreasing homocysteine induced oxidative stress and platelet aggregation [108]. There is also considerable evidence that fish oils and ω3fatty acids [109] could be useful as nutaceuticals to prevent thrombus formation through suppression of thromboxane A2 synthesis and platelet aggregation. Allicin, the aromatic compound from garlic is also a strong inhibitor of platelet aggregation [110]. It is suggested [111] that allicin might initially convert to a compound known as ajoene, which then might attenuate thrombosis through its ability to bind fibrinogen. Another potential compound from garlic is diallyl trisulfide known to inhibit thromboxane synthesis and calcium ion mobilization [112]. It is critical that careful consideration be give to potential interaction of plant derived compounds with other coagulants before their use as dietary supplements since many of them have been shown to interact with Coumadin coagulants possibility through interference with cytochrome P-450 systems [113].

Acknowledgments

The grant support for some of the studies from our laboratory reported in this review from the National Heart Lung and Blood Institute, Bethesda, Maryland and Rio Trading Co. Brighton, UK is gratefully acknowledged.

References

[1] Leys, D. Atherothrombosis: a major health burden. *Cerebrovasc. Dis.* 2011; 11 Suppl. 2: 1-4.
[2] Quist-Paulsen, P., Naess, I. A., Canniegieter, S. C., Romundstad, P. R., Christiansen, S. C., Rosendaal, F. R., Hammerstrom, J. Arterial cardiovascular risk factors and venous thrombosis: results from a population-based prospective study (the HUNT 2). *Haematologica.* 2010; 95: 119-125.
[3] Linden, M. D., Jackson, D. E. Platelets: pleiotropic roles in atherogenesis and Atherothrombosis. *Int. J. Biochem. Cell. Biol.* 2011; 42:1762-1766.
[4] Koenen, R. R., Weber, C. Platelet derived chemokines in vascular remodeling and atherosclerosis. *Semin. Thromb. Hemost.* 2010; 36: 163-169,
[5] Aukrust, A. P., Halvorsen, B., Ueland, T., Michelson, A. E., Skjelland, M., Gullestad, L., Yndestad, A., Otterdal, K. Activated platelets and atherosclerosis. *Expert Rev. Cardiovasc. Ther.* 2010; 8:1297-1307.
[6] Drouet, L., Baldit Solllier, C., Henry, P. *Drugs* 2010; 70 Suppl. 1: 9-14,
[7] Coccheri, S. Antiplatelet drugs-do we need new options? With a reappraisal of direct thromboxane inhibitors. *Drugs.* 2010; 70: 887-908.

[8] Buch, M. H., Prendergast, B. D., Storey, R. F. Anti-platelet therapy and vascular disease: an update. *Ther. Adv. Cardiovasc. Dis.* 2010; 4: 249-275.

[9] Kenet, G., Lutkhoff, L. K., Albisetti, M., Bernard, T., Bonduel, M., et al. Impact of thrombophilia on risk of arterial ischemic stroke or cerebral sinovenous thrombosis in neonates and children: a systematic review and meta-analysis of observational studies. *Circulation.* 2010; 121: 1837-1847.

[10] Muszbek, L., Bereczky, Z., Bagoly, Z., Shemirani, A. H., Katona, E. Factor XIII and atherothrombotic diseases. *Semin. Thromb. Hemost.* 2010; 36:18-33.

[11] Mousa, S. Antithrombotic effects of naturally derived products on coagulation and platelet function. In Anticoagulants, antiplatelets and thrombolytics, Edited by Mousa, S. *Methods in Molecular Biology.* 2010; 663, 229-240,

[12] Bernard, T. J., Manco-Johnson, M. J., Goldenberg, N. A. The role of anatomic factors, thrombophilia and antithrombin therapies in childhood onset arterial ischemic stroke. *Thromb. Res.* 127: 6-12, 2010.

[13] Venter, J. C., Adams, M. D., Myers, E. W., Peter, W., Mural, R. J., et al.: The sequence of human genome. *Science.*2001; 291: 1304-135.

[14] Mälarstig, A., Hamsten, A. Genetics of atherothrombosis and thrombophilia. *Curr. Atheroscler. Rep.* 2010; 12:159-166.

[15] Hansson, G. K..Inflammation, atherosclerosis and coronary artery disease. *N Engl. J. Med.* 2005; 352:1685-1695.

[16] Martinelli, I., Bucciarelli, P., Mannucci, P. M. Thrombotic risk factors: basic pathophysiology. *Crit. Care Med.* 2010; 38 (suppl.): S3-S9.

[17] Esmon, C. T. Basic mechanisms and pathogenesis of venous thrombosis. *Blood Rev.* 2009; 23: 225-229.

[18] Van de Poll, T., de Boer, J. D., Levi, M. The effect of inflammation on coagulation and vice-versa. *Curr. Opin. Infect. Dis.* 2011; 24: 273-278.

[19] Levi, M., van der Poll, T.: Inflammation and coagulation. *Crit. Care Med.* 2010; 38 (suppl. 2): S26-S34.

[20] Naess, I. A., Christiansen, S. C., Romundstad, P., Cannegieter, S. C., Rosendaal, F. R., Hammerstrom, J. Incidence and mortality of venous thrombosis: a population-based study. *J. Thromb. Haemost.* 2007; 5: 692-699.

[21] Rosendaal, F. R., Reitsma, P. H. Genetics of venous thrombosis. *J. Thromb. Haemost.* 2009; 7 (suppl. 1): 301-304.

[22] Lane, D. A., Mannucci, P. M., Bauer, K. A., Bertina, R. M., Bochkov, N. P., Boulyjenko, V., Chandy, M., Dahlbäck, B. Inherited thrombophilia. Part I. *Thrmob. Haemost.* 1996; 76: 651-662.

[23] Soare, A. M., Popa, C. Deficiencies of proteins C, S and antithrombin and factor V Leiden and the risk of ischemic strokes. *J. Med. Life.* 2010;3:235-238.

[24] Bertina, R. M. Molecular risk factors for thrombosis. *Thromb. Haemost.* 1999; 82:601-609.

[25] Spector, E. B., Grody, W. W., Matteson, C. J., Palomaki, G. E., Bellissimo, D. B., Wolff, D. J., Bradley, L. A., Prior, T. W., Feldman, G., Popovich, B. W., Watson, M. S., Richards, C. S. Technical Standards and guidelines: Venous thromboembolism (Factor V Leiden and prothrombin 20210GA testing): A disease-specific supplement to the Standards and guidelines for clinical genetics laboratories. *Genet. Med.* 2005; 7: 444-453.

[26] Ridker, P. M., Miletich, J. P., Hennekens, C. H., Buring, J. E. Ethnic distribution of factor V Leiden in 4047 men and women. Implications for venous thromboembolism screening. *JAMA*; 277:1305-1307.

[27] Emmerich, J., Rosendaal, F. R., Cattaneo, M., Margaglione, M., de Stefano, V., Cumming, T., Arruda, V., Hillarp, A., Reny, J. L. Combined effect of factor V Leiden and prothrombin 20210A on the risk of venous thromboembolism-Pooled analysis of 8 case-control studies including 2310 cases and 3204 controls. *Thromb. Haemost.* 2001; 86:809-813.

[28] Poort, S. R., Rosendaal, F. R., Reitsma, P. H., Bertina, R. M. A common genetic variation in the 3'-untranslated region of the prothrombin gene is associated with elevated plasma prothrombin levels and an increase in venous thrombosis. *Blood.* 1996; 88: 3698-3703.

[29] Bozic-Mijovski, M. Hyperhomocysteinemia and thrombophilia. *Clin. Chem. Lab. Med.* 2010; 48 (suppl.):589-595.

[30] Den Heijer, M., Rosendaal, F. R., Blom, H. J., Gerrits, W. B., Bos, G. M. Hyperhomocysteinemia and venous thrombosis: a meta-analysis. *Thromb. Haemost.*, 1998; 80:874-877.

[31] Den Heijer, M., Brouwer, I. A., Bos, G. M., Blom, H. J., van der Put, N. M., Spaans, A. P., Rosendaal, F. R., Thomas, C. M., Haak, H. L., Wijermans, P. W., Gerrits, W. B. Vitamin supplementation reduces blood homocysteine levels. *Arterioscler. Thromb. Vasc. Biol.* 1998;18:356-361.

[32] Den Heijer, M., Willems, H. P., Blom, H. J., Gerrits, W. B., Cattaneo, M., Eichinger, S., Rosendaal, F. R., Bos, G. M. Homocysteine lowering by B vitamins and the secondary prevention of deep vein thrombosis and pulmonary embolism: a randomized, placebo-controlled, Double-blind *Trial. Blood.* 2007;109: 139-144.

[33] Rees, M. M., Rodgers, G. M. Homocysteinemia: association of a metabolic disorder with vascular disease and thrombosis. *Thromb. Res.* 1993; 71:337-359.

[34] Kang, S.-S., Zhou, J., Wong, P. W., Kowalisyn, J., Strokosch, G. Intermediate homocysteinemia: a thermo-labile variant of methylene-tetra-hydrofolate reductase. *Am. J. Hum. Genet.* 1988; 43: 414-421.

[35] Selhub, J., Jacques, P. F., Wilson, P. W., Rush, D., Rosenberg, I. H. Vitamin status and intake as primary determinants of homocysteinemia in an elderly population. *J. Am. Med. Assoc.*, 1993; 270: 2693-2698.

[36] Cooper, P. C., Coath, F., Daly, M. E., Makris, M. The phenotypic and genetic assessment of antithrombin deficiency. *Int. J. Lab. Hem.* 2011; 33:227-237.

[37] Patnaik, M. M., Moll, S. Inherited antithrombin deficiency: a review. *Haemophilia.* 2008;14:1229-1239.

[38] Ishiguro, K., Kojima, T., Kadomatsu, K., Nakayama, Y., Takagi, A., Suzuki, M., Takeda, N., Ito, M., Yamamoto, K., Matsushita, T., Kusugami, K., Muramatsu, T., Saito, H. Complete antithrombin deficiency in mice results in embryonic lethality. *J. Clin. Invest.* 2000;106: 873-878.

[39] Castoldi, E., Rosing, J. APC resistance: biological basis and acquired influences. *J. Thromb. Haemost.* 2009; 8: 445-453.

[40] Segers, O., Castoldi, E. Factor V Leiden and activated protein C resistance. *Adv. Clin. Chem.* 2009;49: 121-157.

[41] Griffin, J. H., Fernandez, J. Á., Gale, A. J., Mosnier, L. O. Activated protein C. *J. Thromb. Haemost.* 2007; 5 (suppl. 1): 73-80.

[42] Navarro, S., Bonet, E., Estellés, A., Montes, R., Hermida, J., Martos, L., España, F., Medina, P. The endothelial cell protein C receptor: its role in thrombosis. *Thromb. Res.* 2011;128: 410-416.

[43] Dahlbäck, B., Villoutreiv, B. O. The anticoagulant protein C pathway. *FEBS Lett.* 579: 2005; 3310-3316.

[44] Norstrom, E. A., Steen, M., Tran, S., Dahlbäck, B. Importance of proteins and phospholipid for activated protein C-mediated cleavages in factor Va. *J. Biol. Chem.* 2003;278: 24904-24911.

[45] Dawson, S., Wiman, B., Hamstem, A., Green, F., Humphries, S., Henney, A. M. The two allele sequences of a common polymorphism in the promoter of the plasminogen activator inhibitor-1 (PAI-1) gene respond differently to interleukin-1 in HepG2 cells. *J. Biol. Chem.* 1993; 268:10739-10745.

[46] Lane, D. A., Grant, P. J. Role of hemostatic gene polymorphisms in venous and arterial thrombotic disease. *Blood.* 2000; 95: 1517-1532.

[47] Smith, E. B. Haemostatic risk factors for cardiovascular diseases. *Eur. Heart J.* 1998;19:39-43.

[48] Tsantes, A. E., Nikolopoulos, G. K., Bagos, P. G., Bonovas, S., Kopterides, P., Vaiopoulos, G. The effect of the plasminogen activator inhibitor-1 4G/5G polymorphism on the thrombotic risk. *Thromb. Res.* 2008; 122:736-742.

[49] Tanis, B. C., Rosendaal, F. R. Venous and arterial thrombosis during oral contraceptive use: risk and risk factors. *Semi. Vasc. Med.* 2003; 3 69-84.

[50] Rosendaal, F. R., Van Hylckama, V. A., Tanis, B. C., Helmerhorst, F. M. Estrogens, progestogens and thrombosis. *J. Thromb. Haemost.* 2003;1:1371-1380.

[51] Mohllajee, A. P., Curtis, K. M., Martins, S. L., Peterson, H. B. Does use of hormonal contraceptives among women with thrombogenic mutations increase their risk of venous thromboembolism? A systematic review. *Contraception.* 2005; 73;166-178.

[52] Vaya, A., Mira, Y., Mateo, J., et al. Prothrombin G20210A mutation and oral contraceptive use increase upper-extremity deep vein thrombotic risk. *Thromb. Haemost.* 2003; 89:452-457.

[53] Legnani, C., Cini, M., Cosmi, B., et al. Risk of deep vein thrombosis: interaction between oral contraceptives and higher factor VIII levels. *Haematologia.* 2004; 89:1347-1351.

[54] Kemmeren, J. M., Algra, A., Meijers, J. C., et al. Effect of second and third generation oral contraceptives on protein C system in the absence or presence of the factor V Leiden mutation: a randomized trial. *Blood.* 2004;104: 927-933.

[55] Heinemann, L. A., Lewis, M. A., Spitzer, W. O., et al. Thromboembolic stroke in young women. A European case-controlled study on oral contraceptives. translational research group on oral contraceptives and the health of young women. *Contraception.* 1998;57: 29-37.

[56] Wassertheil-Smoller, S., Hendrix, S. L., Limacher, M., et al. Effect of estrogen plus progestin on stroke in postmenopausal women: the Women's Health Initiative: a randomized trial. *JAMA.* 2003; 289:271-279.

[57] Cushman, M., Kuller, L. H., Prentice, R., et al. Estrogen plus progestin and risk of venous thrombosis. *JAMA.* 2004; 292:1573-1580.

[58] Smith, N. L., Heckbert, S. R., Lemaitre, R. N., et al. Esterified estrogens and conjugated equine estrogens and risk of venous thrombosis. *JAMA*. 2004; 292:1581-1587.
[59] Hulley, S., Grady, D., Bush, T. Heart and Estrogen/progestin Replacement Study (HERS). Randomized trial of estrogen plus progestin for secondary prevention of coronary heart disease in postmenopausal women. *JAMA*. 1998;280: 605-612.
[60] Herrington, D. M., Reboussin, D. M., et.al. Effects of estrogen replacement on the progression of coronary artery atherosclerosis. *N Eng. J. Med.* 2000; 343:522-529.
[61] Writing Group for the Women's Health Initiative Investigators. Risks and benefits of estrogen plus progestin in healthy postmenopausal women. Principal results from the Women's Health Initiative randomized Controlled Trial. *JAMA*. 2002;288:321-333.
[62] Glueck, C. J., Wang, P., Fontaine, R. N., Tracy, T., et al. Effect of exogenous estrogen on atherothrombotic vascular disease risk related to the presence or absence of the factor V Leiden mutation. *Amer. J. Cardiol.* 1999;84:549-554.
[63] Glueck, C. J., Wang, P., Fontaine, R. N., Sieve-Smith, L., Lang, J. E. Interaction of estrogen replacement therapy with the thrombophilic 20210 G/A prothrombin gene mutation for atherothrombotic vascular disease: a cross-sectional study of 275 hyperlipidemic women. *Metabolism*. 2001;50: 360-365.
[64] Psaty, B. M., Smith, N. L., Lemaitre, R. N., Vos, H. L., Heckbert, S. R., LaCoix, A., Rosendaal, F. R. Hormone replacement therapy, prothrombotic mutations and the risk of incident non-fatal myocardial infarction in postmenopausal women. *JAMA*. 2001; 285: 906-913.
[65] Gardemann, A., Lohre, J., Katz, N., Tillmans, H., Hehrlein, F. W., Haberbosch, W. The 4G/4G genotype of the plasminogen activator inhibitor 4G/5G gene polymorphism is associated with coronary atherosclerosis in patients at high risk for the disease. *Thromb. Heamost.* 1999;82: 1121-1126.
[66] Botto, N., Maffei, S., Manfredl, S., Colombo, M. G., Mazzone, A. M., Andreassi, M. G. Prothrombin mutations, family history, and the risk of thrombosis in postmenopausal women: Implications for hormone replacement therapy. *Climateric*. 2011; 14: 25-30.
[67] Dentali, F., Squizzato, A., Ageno, W. The metabolic syndrome as a risk factor for venous and arterial thrombosis. *Semin. Thromb. Hemost.* 2009; 35:451-457.
[68] Subbiah, M. T. R. Cardiovascular risk associated with estrogen replacement therapy: some critical questions about the type of estrogen used in most clinical trials. *Atherosclerosis*. 2002; 173, 373-374.
[69] Lanas, A., Garcia-Rodriguez, L. A., Arroyo, M. T., et al. Risk of upper gastrointestinal bleeding associated with selective COX-2 inhibitors, traditional NSAIDs, aspirin and combinations. *Gut*. 2006; 55:1731-1738.
[70] Zuern, C. S., Lindemann, S., Gawaz, M. Platelet function and response to aspirin: gender-specific features and implications for female thrombotic risk and management. *Semin. Thromb. Hemost*. 2009; 35:295-306.
[71] Erickson, H. T., Correa, M. P. F., Escobar, J. R. Guarana. *Econ. Botany*. 1984; 38:273-286.
[72] Billiard, F., Martello, A., Valle, M. G. HPLC determination of caffeine and theophylline in Paullina cupana Knuth (guarana) and cola spp. samples. Z. Lebanese. *Enters. Forrsch.* 1985; 180, 398-401.
[73] Bempong, D., Houghton, P. J. Dissolution and absorption of caffeine from guarana. *J. Pharm. Pharmacol.* 1992; 44:769-771.

[74] Bempong, D., Houghton, P. J., Steadman, K. The caffeine content of guarana. *J. Pharm. Pharmacol.* 1991;43 (suppl.),125.

[75] Marx, F. Analysis of guarana seeds. II. Studies on the composition of tannin fraction. *J. Lebensm Unters Forsch.* 1990;190, 429-431.

[76] Subbiah, M. T. R. Guarana consumption. A review of health benefits and risks. *Altern. Compliment Ther.* 2005;4, 212-213.

[77] Haskell, C. K., Kennedy, D. O., Wesnes, K. A., Milne, A. L., Scholey, A. B. A double-blind, placebo-controlled, multi-dose evaluation of the acute behavioral effects of guarana in humans. *J. Psychopharmacol.* 2007; 21: 65-70.

[78] Berube-Parent, S., Pelletier, C., Dore, J., Trembley, A. Effects of encapsulated green tea and guarana extracts containing a mixture of epigallocatechin-3-gallate and caffeine on 24 hr energy expenditure and fat oxidation in men. *Brit. J. Nutr.* 2005;94:432-436.

[79] Opala, T., Rzymski, P., Pischel, I., Wilczak, M., Wozniak, J. Efficacy of 12 weeks supplementation of a botanical extract based weight loss formula on body weight, body composition and blood chemistry in healthy overweight subjects: a randomized double-blind placebo controlled clinical trial. *Eur. J. Med. Res.* 2006;11:342-350.

[80] Bydlowski, S. P.,Yunker, R. L., Subbiah, M. T. R. A novel property of aqueous guarana extracts (Paullina cupana): inhibition of platelet aggregation. *Bras. J. Med. Biol. Res.* 1988; 21,535-538.

[81] Bydlowski, S. P., D'Amico, E. A., Chamone, D. A. An aqueous extract of guarana (Paullinia cupana) decreases platelet thromboxane synthesis. *Bras. J. Med. Biol. Res.* 1991;24,421-424.

[82] Subbiah, M. T. R., Yunker, R. L. Studies on the nature of anti-platelet aggregatory factors in the seeds of the Amazonian herb guarana (Paullinia cupana). *Int. J. Vitam. Nutr. Res.* 2008;78: 96-101.

[83] Oliveira, C. H., Moraes, M. E., Moraes, M. O., Bezerra, F. A., Abib, E., De Nucci, G. Clinical toxicology study of an herbal medicinal extract of Paullinia cupana, Trichilia catigua, Ptycopetalum olacoides and Zingiber officinale in healthy volunteers. *Phytother. Res.* 2005;19:54-57.

[84] Haller, C. A, Jacob, P., Benowitz, N. L. Pharmacology of ephedra alkaloids and caffeine after single dose dietary supplement use. *Clin. Pharmacol. Ther.* 2002;71:421-432.

[85] Baghkhani, L., Jafari, M. Cardiovascular adverse reactions associated with guarana: is this a causal effect? *J. Herbal Pharmacology.* 2002; 2: 57-61.

[86] Gorlach, A. Redox regulation of the coagulation cascade. *Antioxid. Redox. Signal.* 2005; 7: 1398-1404.

[87] Essex, D. W. Redox control of platelet function. Antioxid Redox Signal. 2009;11: 1191-1225.

[88] Coulon, L., Calzada, C., Moulin, P., et al. Activation of p38 mitogen-activated protein hydroperoxide-stressed platelets. *Free Radic. Biol. Med.* 2003;35: 616-625.

[89] Seno, T., Inooue, N., Gao, D., et al. Involvement of NADH/NADPH oxidase in human platelet ROS production. *Thromb. Res.* 2001;103:399-409.

[90] Cieslar, P., Masova, L., Scheiner, T., et al. Oxidative stress and platelet function in multiple myeloma and renal insufficiency: clinical relations of different states. *Thromb. Res.* 2002; 105:277-283.

[91] Freeman, J. E., Keaney, J. F., Jr. Vitamin E inhibition of platelet aggregation is independent of antioxidant activity. *J. Nutr.* 2001;131:374s-377s.

[92] Glynn, R. J., Ridker, P. M., Goldhaber, S. Z., et al. Effects of random allocation to vitamin E supplementation on the occurrence of venous thromboembolism: report from the Women's Health Study. *Circulation.* 2007;116:1497-1503.

[93] Loscalzo, J., Welch, G. Nitric oxide and its role in the cardiovascular system. *Prog. Cardiovasc. Dis.* 1995;38:87-104.

[94] Loscalzo, J. Nitric oxide insufficiency, platelet activation and arterial thrombosis. *Circ. Res.* 2001;88:756-762.

[95 Freedman, J. E., Frei, B., Welch, G. N., et al. Glutathione peroxidase potentiates the inhibition of platelet function by s-nitrothiols. *J. Clin. Invest.* 1995; 96:394-400.

[96] Abdulah, R., Katsuya, Y., Kobayashu, K., et al. Effect of sodium selenite supplementation on the levels of prostacyclin I and thromboxane A2 in humans. *Thromb. Res.* 119:305-310, 2007

[97] Chakrabarthi, S., Freedman, J. E. Review: Nutriceuticals as antithrombotic agents. *Cardiovasc. Therapeutics.* 2010; 28:227-235.

[98] Freedman, J. E., Parker, C., Li, L., et al. Select flavonoids and whole juice from purple grapes inhibit platelet function and enhance nitric oxide release. *Circulation.* 2001; 103: 2792-2798.

[99] Vitseva, O., Varghese, S., Chakrabarthi, S., et al. Grape seed and skin extracts inhibit platelet function and release of reactive oxygen intermediates. *J. Cardiovasc. Pharmacol.* 2005; 46:445-451.

[100] Polette, A., Lemaitre, D., Lagarde, M., et.al. N-3 fatty acid-induced lipid peroxidation is prevented by catechins. *Thromb. Haemost.* 1996; 75:945-949.

[101] Pignatelli, P., Lenti, L., Pulcinelli, F. M., et al. Red and white wine differently affect collagen induced platelet aggregation. *Pathophysiol. Haemost. Thromb.* 2002; 32:356-358.

[102] Pignatelli, P., Di Santo, S., Buchetti, B., et al. Polyphenols enhance platelet nitric oxide by inhibiting protein kinase C dependant NADPH oxidase activation: effect on platelet recruitment. *FASEB J.* 2006; 20:1082-1089.

[103] Kang, W. S., Chung, K. H., Chung, J. H., Lee, J. Y., Park, J. B., Zhang, Y. H., Yoo, H. S., Yun, Y. P. Antiplatelet activity of green tea catechins is mediated by inhibition of cytoplasmic calcium release. *J. Cardiovasc. Pharmacol.* 2001; 38:875-884.

[104] Hodgson, J. M., Puddey, I. B., Burke, V., Beilin, L. J., Mori, T. A., Cahn, S. Y. Acute effects of ingested black tea on postprandial platelet aggregation in human subjects. *Brit. J. Nutr.* 2002; 87:141-145.

[105] Kondo, K., Suzuki, Y., Ikeda, Y., Umemura, K. Genistein, an isoflavone included in soy, inhibits thrombotic vessel occlusion in the mouse femoral artery and in vitro platelet aggregation. *Eur. J. Pharmacol.* 2002; 455:53-57.

[106] McNicol, A. The effects of genistein on platelet function are due to thromboxane receptor antagonism rather than inhibition of tyrosine kinase. *Prostaglandins Leukot Essen Fatty Acids.* 1993; 48:379-384.

[107] Tan, L., Kowalska, M. A., Bomo, G. M., Lopez, J. A., Darzynkiewicz, Z., Niewiarowski, S. Identification and characterization of endothelial glycoprotein 1b using viper venom proteins modulating cell adhesion. *Blood*; 93:2605-2616.

[108] Casais, P., Alberto, M. F., Salviu, M. J., et al. Pilot study of homocysteine, and cysteine in patients with thrombosis in different vascular sites. Epidemiology and response to folate. *Thromb. Res*. 2009;123:592-596.

[109] Braden, G. A., Knapp, H. R., Fitzgerlansd, D. J., et al. Dietary fish oil accelerates the response to coronary thrombolysis with tissue-type plasminogen activator. Evidence for a modest platelet inhibitory effect in vivo. *Circulation*. 1990;82:178-187.

[110] Makheja, A. N., Bailey, J. M. Antiplatelet constituents of garlic and onion. *Agents Actions*. 1990; 29:360-363.

[111] Srivastava, K. G., Tyagi, O. D. Effect of garlic derived principle (ajeone) on aggregation and arachidonic acid metabolism in human blood platelets. *Prostaglandins Leukot Essen Fatty Acids*. 1993;49: 587-595.

[112] Qi, R., Liao, F., Inoue, K., Yatomi, Y., Sato, K., Ozaki, Y. Inhibition of diallyl trisulfide, a garlic component, of intracellular Ca(2+) mobilization without affecting inositol-1,4,5-triphosphate formation in activated platelets. *Biochem. Pharmacol*. 2000;60:1475-1483.

[113] Mousa, S. A. Antithrombotic effects of naturally derived products on coagulation and platelet function. In: Anticagulants, Antiplatelets and Thrombolytics. Mousa, S. A Editor. *Methods in Molecular Biology*; 663:229-240.

Nutrigenomics and Cosmeceutical Development

In: Nutrigenomics
Editor: M. T. Ravi Subbiah

ISBN: 978-1-62417-051-5

Chapter VII

Nutritional Genomics in Skin Health: Application to the Development of Nutriceuticals and Cosmeceuticals

Xue-Gang Xu, Hong-Duo Chen and Xing-Hua Gao*
Department of Dermatology, No.1 Hospital of
China Medical University, Shenyang, China

Abstract

Nutritional genomics, the application of high throughput genomics tools in nutrition, has tremendous potential to change the future of dietary guidelines and personal recommendations. Nutritional genomics includes nutrigenomics and nutrigenetics. It is highly likely that during the next decade the nutritional supplement and functional food industries will continue to have a robust growth in response to advances in nutritional genomics research and its applications. The research and development of cosmeceuticals is booming in recent years. Many substances, either from botanical, animal, or chemically synthesized sources, are tested or investigated as the active ingredients in cosmeceuticals. Nutritional genomics will promote an increased understanding of how cosmeceuticals/nutriceuticals influence metabolic pathways and homeostatic control. Eventually, nutrigenomics/nutrigenetics may lead to personalized cosmeceuticals/ nutriceuticals recommendations for skin health.

Keywords: nutrigenomics, nutraceuticals, cosmeceuticals, skin health, dermagenetics, pigment lightening agents, hydrquinone, peptides, antioxidants, growth factors, botanicals, retinoids

* Department of Dermatology, No.1 Hospital of China Medical University. 155 North Nanjing Street, Shenyang, 110001, China.

Introduction

Traditional medical practices, for example, Chinese medicine and Ayurveda (Indian traditional medicine), have utilized the medicinal properties of food for thousands of years. Traditional herbal medicine formulas are not designed to treat symptoms of a specific illness; rather, they are tailored specifically to the individual according to the complex principles of oriental medicine [1]. Generally, the birth of nutrition as a 'science' can be placed at around 1785, coincident with the advent of the 'chemical revolution' [2]. During the past 220 years of nutrition research, we have experienced the chemical revolution and we have traversed the golden age of nutrition ('vitamin era' 1912-1940s). Emphasis has transferred from deficiency-related diseases to over-nutrition (chronic diseases) [2]. Recently, the major practical impact of nutrition research on public health resides in the formulation of optimal dietary recommendations that are aimed at preventing diseases and, increasingly, at promoting optimal health and aging [2]. Now is the time for a 'genomic revolution'.

Nutrigenomics and Nutrigenetics

The completion of the Human Genome Project has sparked a great deal of research on individual variation in gene sequences, particularly in single nucleotide polymorphisms (SNPs), their role in chronic diseases, and in predicting individual responses to drugs, to name a few. Recent advances in human genetics empower researchers to discover the genetics of many common disorders.

The advent of genomics, interpreted broadly as a suite of high throughput technologies for the generation, processing, and application of scientific information about the composition and functions of genomes, has created unprecedented opportunities for increasing our understanding of how nutrients modulate gene and protein expression and ultimately influence cellular and organismal metabolism [3]. Nutritional genomics, the junction between health, diet, and genomics, can be seen as the combination of molecular nutrition and genomics. The diverse tissue and organ-specific effects of bioactive dietary components include gene-expression patterns (transcriptome); organization of the chromatin (epigenome); protein-expression patterns, including post-translational modifications (proteome); as well as metabolite profiles (metabolome) [3].

Nutritional genomics includes [1] nutrigenomics: the study of interaction of dietary components with the genome; and [2] nutrigenetics: understanding the gene-based differences in response to dietary components and developing nutraceuticals that are most compatible with health for individuals based on their genetic makeup [4]. Nutrigenomics explores the effects of nutrients on the genome, proteome and metabolome. Nutrigenetics elucidates the effect of genetic variation on the interaction between diet and disease [2]. Eventually, nutritional genomics will lead to evidence-based dietary intervention strategies for restoring health and fitness and for preventing diet-related disease [3].

Why is there a growing realization on nutritional genomics? First, the completion of several large genome projects has markedly altered the research agenda by drawing attention to the importance of genes in human nutrition, and has provided a wealth of new genetic information to be explored. Second, there has been a growing recognition that micronutrients and macronutrients can be potent dietary signals that influence the metabolic programming of cells and have an important role in the control of homeostasis. Third, nutrition researchers have increasingly started to recognize that genetic predisposition can be an important contributor to the main causes of mortality that are linked to diet, such as cardiovascular disease, diabetes and cancers [5, 6]. The challenge for the next decade is to identify nutrient-influenced molecular pathways and determine the down-stream effects of specific nutrients. Nutrigenomics can assist in this identification because it allows the genome-wide characterization of genes, the expression of which is influenced by nutrients [5].

An important challenge in nutrition research is the complexity and variability of nutrition and foods. The body has to handle a large number of different nutrients and other food components and nutrient concentrations can be high without reaching toxic levels. Each nutrient can also have numerous targets with different affinities and specificities. This situation contrasts starkly with pharmacology, where single agents are used at low concentrations and act with a relatively high affinity and selectivity for a very limited number of biological targets. The challenge of nutrigenomics research is to break down the important but complex research problems into small feasible projects that can be handled by normal-sized research groups [3]. A possible strategy is to use methods that are well established in medical or pharmacological research but are rather new to nutrition research. For example, analogous to pharmacology, nutrients or dietary metabolites can be viewed as signaling molecules that are recognized by specialized cellular-sensing mechanisms. The information that allows nutrients to activate specific signaling pathways is contained within their molecular structure. Minor changes in structure may have a profound influence on which sensor pathways are activated. The challenge ahead is to further identify the molecular pathways that are influenced by individual nutrients and to determine the downstream effects of this regulation. Nutrigenomics can greatly assist in this identification because it allows the genome-wide characterization of nutritional target genes. A single nutrigenomics experiment can generate an enormous amount of data. Subtle changes in gene expression, even at the single-cell level, can now be measured by quantitative techniques such as real-time PCR and high-density microarray analysis. The latter allows the entire nutrition-relevant transcriptome to be studied simultaneously [5]. With this type of information, researchers can comprehensively understand how nutrients act and explain how diet has such an impressive effect on health and disease, which is now widely acknowledged. Ultimately, nutrigenomics research will lead to development of evidence-based healthful food and lifestyle advice and dietary interventions for humans [3].

Nutrigenetics has been used for decades in certain rare monogenic diseases such as phenylketonuria, and it has the potential to provide a basis for personalized dietary recommendations based on the individual's genetic makeup in order to prevent common multifactorial disorders decades before their clinical manifestation [2]. For example, the APOA1 gene codes for the major protein in HDL cholesterol and plays a central role in lipid metabolism. A particular variant of this gene in which A (adenosine) is substituted for G (Guanosine) at position -75 affects how the person responds to PUFAs. When researchers evaluated the effect of PUFA intake on HDL cholesterol levels in different populations, a significant sex difference was noted. Women with the GG allele had higher HDL cholesterol concentrations in response to low levels of PUFA (<4%), whereas GA women had higher HDL cholesterol concentrations only when PUFA intake was >8% of total energy. These same genetic-dietary interactions were not significant in men. This type of knowledge is helpful in allowing clinicians to target their recommendations to those who will benefit from a diet rich in PUFAs and those who will not or may even be harmed by such a recommendation [7]. The food industry also recognizes the need for nutrigenomics as a basis for developing the concept of "personalized diets" for identifying molecular biomarkers or new bioactive food ingredients, and for validating the effectiveness of these bioactive ingredients as functional food components or nutraceuticals.

Active Ingredients in Cosmeceuticals/Nutriceuticals

Former researches of nutritional genomics were mainly focused on chronic diseases such as cardiovascular disease, diabetes type II and cancers, but only quite few studies recognized the application of nutritional genomics in the development of cosmeceuticals and nutraceuticals for skin health [8]. The aging population and a desire to maintain a youthful appearance have propelled the cosmeceuticals markets. Today it is a commonplace for consumers to use cosmetics and functional food to support their beauty and general health, protecting the body from environment hostile for the human wellness. Thus, products that claim to produce skin health benefits from the outside as well as the inside are beginning to appear on the market. When evaluating the application of nutritional genomics to skin health, we have to pay more attention to the use of nutrigenomics/ nutrigenetics in the development of cosmeceuticals and nutraceuticals.

Cosmeceutical is a category of cosmetic products claimed to have biologically active ingredients with medical or drug-like benefits. The term is attributed to Dr. Albert Kligman, who identified a hybrid category of products lying on the spectrum between drugs and cosmetics more than 25 years ago. It is now commonly used to describe a cosmetic product that exerts a pharmaceutical therapeutic benefit but not necessarily a biologic therapeutic benefit [9]. Cosmeceuticals are intended to carry out their functions as protection, whitening, tanning, antiwrinkling, deodorants, antiaging, and nail and hair care [10]. The interactions between cosmeceuticals and skin are complex, depending on the specific composites in cosmeceutical products, condition of the skin or general health status of a subject, and the environment where the action occurs [10]. Cosmeceuticals are basically functional cosmetics; the ingredients included in formulations must come from a list of raw materials that are generally recognized as safe [11]. Cosmeceuticals are mainly divided into the following broad categories: pigment lightening agents, antioxidants, growth factors, peptides, botanicals [9]. Before we probe into the application of nutrigenomics/ nutrigenetics in cosmeceuticals and nutriceuticals, we first review some of the cosmeceuticals/ nutriceuticals with different categories of functions, and mainly focus on their biologically active ingredients.

Pigment Lightening Agents

Protease-activated receptor 2 (PAR-2) is a seven transmembrane G-protein-coupled receptor. Inhibition of PAR-2 activation by synthetic serine protease inhibitors results in depigmentation. These inhibitors act through inhibition of PAR-2 cleavage affecting cytoskeletal and cell surface organization, thus preventing keratinocyte-melanocyte contact that results in reducing melanosome phagocytosis by keratinocytes. Other non-bleaching agents such as vitamins C and E, and a polyphenol ellagic acid have been reported to decrease tyrosinase activity, which is the key enzyme in the synthesis of melanin pigment [10, 12].

Hydroquinone is well known for its suppressive effect on melanin synthesis. Topical hydroquinone in 2% to 4% concentrations, alone or in combination with tretinoin 0.05% to 0.1%, has been successfully used for years [13]. Derivatives of hydroquinone have also shown some promises for lightening skin. Arbutin contains a form of hydroquinone derived from the leaves of bearberry, cranberry, mulberry, or blueberry shrubs, and in most types of pears [10].

Kojic acid, a natural substance produced by fungi or bacteria, such as *Aspergillus*, *Penicillium*, or *Acetobacter spp*, is present in traditional Japanese fermented foods. There is convincing evidence showing kojic acid to be effective for inhibiting melanin production [9].

Glabridin is licorice extract and can inhibit tyrosinase activity. It also has anti-inflammatory properties through cyclooxygenase inhibition.

Ellagic acid is a polyphenol found in pomegranates. It inhibits tyrosinase by chelating copper at the active center of this enzyme. It may selectively inhibit melanin synthesis only in UV-activated melanocytes.

Many of the cosmeceuticals already described also have pigment-lightening effects. Vitamin C and E decrease tyrosinase activity. Pycnogenol decreases UV-induced pigmentation. Niacinamide inhibits transfer of melanosomes to epidermal keratinocytes. soybean trypsin inhibitor (STI) and Bowman-Birk inhibitor (BBI) in soy can reduce melanin transfer [9].

Antioxidants

Free radicals are highly reactive molecules with unpaired electrons, and they can cause cellular damage to cell membranes, lipids, proteins, and DNA. Antioxidants comprised a group of diverse molecules including, but not limited to, vitamins (A, B, C, E), alpha lipoic acid (ALA), Coenzyme Q-10, idebenone, polyphenols, and kinetin.

Vitamin A, or retinol, is an antioxidant member of the retinoid family, which includes tretinoin. Application of retinol can increase epidermal water content, epidermal hyperplasia, and cell renewal while enhancing collagen synthesis [14]. The vitamin B complex includes niacinamide (Vitamin B_3) and panthenol (provitamin B_5). Topical niacinamide has been shown to have anti-tumor activity on keratinocytes and to suppress UVB-induced photo-carcinogenesis in vitro. Niacinamide improves the lipid barrier component of the epidermis by increasing ceramide and free fatty acids, as well as the epidermal barrier proteins such as keratin, filaggrin, and involucrin. Vitamin C (ascorbic acid) increases mRNA expression for type I collagen, inhibits elastin biosynthesis, inhibits tyrosinase and decreases melano-genesis, which causes the improvement of photodamage and clinical lightening of melasma and lentigines [15]. Vitamin E decreases sunburn cells after UV exposure, neutralizes free radicals, and also acts as a humectant.

Panthenol is a water-soluble cosmeceutical that easily penetrates the stratum corneum and is a humectant. It promotes lipid synthesis, fibroblast proliferation and epidermal re-epithelialization.

Alpha lipoic acid (ALA) is a lipoamide synthesized in the mitochondria of plants and animals. ALA regenerates endogenous antioxidants such as Vitamins C and E, glutathione, and ubiquinol. ALA is both water and lipid soluble, allowing it to penetrate lipophilic cell membranes and enter the aqueous intracellular matrix. The molecule prevents lipid peroxidation, has anti-inflammatory properties, and acts as an exfoliant [16].

CoQ-10 is a fat-soluble antioxidant. It acts by downregulating MMPs. It also inhibits lipid peroxidation. CoQ-10 can decrease periorbital wrinkles. Idebenone is a more potent synthetic analog of CoQ-10. In an in vitro comparison of antioxidants (including idebenone, tocopherol, kinetin, ubiquinone, ascorbic acid, and lipoic acid) idebenone scored highest using 5 different assays to measure anti-inflammatory properties, photoprotective effects, and prevention of UV immunosuppression. It may repair mitochondrial DNA and decrease nuclear thymine dimer photoproducts.

Polyphenols are plant-derived antioxidants that have anti-inflammatory, photoprotective, and anticarcinogenic properties. Flavonoids are a subgroup of polyphenols. They include grape seed extracts, green tea extracts, and soy isoflavones. Grape seed extracts can induce vascular endothelial growth factor expression on keratinocytes to enhance dermal wound healing. Green tea extracts such as epigallocatechin 3-allate have been shown to decrease levels of UVB damage, DNA damage, sunburn, and erythema [17]. Fresh soy milk contains 2 serine protease inhibitors: soybean trypsin inhibitor (STI) and Bowman-Birk inhibitor (BBI). STI inhibits melanosome transfer to keratinocytes, which can lighten the skin. BBI inhibits hair growth presumably as an inhibitor of ornithine decarboxylase.

Kinetin is a plant-derived synthetic growth hormone that has significant antioxidant properties and photo-protective effects, which improves the appearance of fine wrinkles, color, texture, and blotchiness.

Peptides

Peptides are short amino acid sequences that are components of larger proteins, such as collagen. Copper is an essential cofactor for enzyme superoxide dismutase, collagen and elastin formation, down regulates MMPs, and reduces the activity of collagenase. Peptides may be used to stabilize and deliver copper into cells. The tripeptide glycyl-L-histidyl-L-lysine-copper complex is used as a copper vehicle.

As a cosmeceutical, copper peptide is thought to improve skin firmness and texture, fine lines, and hyperpigmentation [18].

Pal-KTTKS is a procollagen pentapeptide fragment that stimulates collagens I and III and fibronectin production. Conjugation of palmitic acid to pal-KTTKS (Palmitoyl-Pal-KTTKS) imparts lipophilic properties that facilitate stratum corneum penetration.

Acetyl hexapeptide -3 (argireline) is a synthetic peptide that inhibits SNARE complex formation, which is involved in synaptic vesicle exocytosis in vitro. Theoretically, argireline may mimic the effects of botulinum toxin injections. Although a promising idea, it has yet to be proven whether this topically applied agent can penetrate to the level of the neuromuscular junction.

Dimethylaminoethanol is a membrane stabilizer that purportedly improves facial muscle tone by releasing acetylcholine. It has been used as a homeopathic oral supplement to enhance mental and physical performance. And 3% topical dimethylaminoethanol gel showed some evidence of skin firming in perioral, periorbital, and cheek skin.

Growth Factors

Growth factors comprise a large group of regulatory proteins that attach to cell surface receptors to mediate inter- and intracellular signaling pathways. Growth factors relevant to wound healing may induce new collagen, elastin, and glycosaminoglycan formation and mediate angiogenesis. One human growth factor presently used in cosmeceuticals is TGF-β1, which is derived from cultured neonatal foreskin fibroblasts.

Botanicals

The search for active ingredients has led to the gathering of flowers, seeds, roots, leaves, twigs, and berries from plants and so on. New botanical material received in the laboratory was purified first. Purified fraction exposed to gene array chip for upregulation or downregulation of key events in cellular oxidation, inflammation or irritation. Then they were studied in culture cells, mouse models and clinical studies. Successful studies pave the way for successful introduction into the marketplace through ingredient licensing arrangements [19].

Resveratrol, found in the skin of red grapes and other fruits, is a stilbenoid. It has also been produced by chemical synthesis or by biotechnological synthesis (metabolic engineered microorganisms) and is sold as a nutritional supplement and active ingredient of cosmeceuticals.

It is well known for its potent ability of anti-inflammation, antioxidant, inhibition of platelet aggregation and cancer cell growth, blood-sugar-lowering and other beneficial cardiovascular effects.

Lycopene is a carotenoid that has potential antioxidant and anticarcinogenic effects both orally and topically. Lycopene is thought to prevent prostate cancer when ingested orally. On the skin, it is protective against UVB photodamage by preventing UVB-induced apoptosis.

Pycnogenol has antimicrobial, anti-inflammatory, antioxidant, and anti-carcinogenic properties. It can accelerate wound healing, reduce scar formation, stabilize elastin fibers, decrease erythema after UV radiation and improve UV-induced pigmentation. It reduces the vitamin C radical which, in turn, regenerates vitamin E, thus recycling the endogenous antioxidant enzyme system [20].

Silymarin is a polyphenolic flavonoid that inhibits UVB-induced sunburn, apoptotic cell formation, and edema. Silymarin also reduces pyrimidine dimer formation, has anticarcinogenic effects, and inhibits COX-2 and IL-1α. Quercetin is also a flavonoid that has antioxidant, anti-inflammatory, anticarcinogenic, and antihistamine effects.

Allantoin is derived from the comfrey root and is also commercially manufactured now. It is an antioxidant, anti-inflammatory agent, and a soothing keratolytic agent. It induces cell proliferation, promotes repair of photodamage, and reduces UV-induced inflammation.

Chamomile and Aloe Vera

Feverfew inhibits NF-κB-dependent gene transcription in a manner similar to that of corticosteroids. Moisturizers that contain feverfew showed significantly improvements in erythema and tactile roughness.

Curcumin has antioxidant, anticarcinogenic, antimicrobial, and anti-inflammatory effects through the inhibition of lipoxygenase and cyclo-oxygenase, prostaglandins, various proinflammatory cytokines, collagenase, elastase, and hylauronidase.

Sandalwood oil is an extract from the dark rough bark and scented nature wood of *Santalum album L*, which is a potent antiseptic, astringent, anti-inflammatory agent. It stimulates cellular regeneration and assists in healing scars, wounds, wrinkles, acne, and other skin conditions. Topical application of 5% sandalwood oil could attenuate UVB-induced erythema, discomfort and sunburn cell formation. It could inhibit UVB-induced upregulation of MMP-1 and MMP-9 mRNA and protein, inhibit UVB-induced increase of IL-10 mRNA, and reduce p53 expression.

Folate is essential in the cell for the synthesis of dTTP from dUMP. Insufficient synthesis of dTTP from dUMP could result in shortened telomere length. Under low folate conditions the thymidine-rich telomere repeat sequence maybe particularly prone to DNA breaks, leading to telomere shortening thus increasing chromosomal instability and DNA damage. Folate and other methyl donors such as vitamin B_{12}, choline and methionine also play a critical role in maintenance of methylation of cytosine which, apart from its importance for transcriptional control of gene expression, determines the structural stability of important regions of the chromosomes such as the centromeres and the sub-telomeric DNA [21].

Retinoids

The first ingredients that met the cosmeceutical criteria were topical retinoids. As an example, we elaborate on the mechanisms of retinoids as a cosmeceuticals. Retinoids are natural and synthetic vitamin A deriatives. The eligible candidates in the group of topical cosmeceutical retinoids are: retinyl esters, retinol, retinaldehyde, and the group of oxoretinoids. They are lipophilic molecules and easily penetrate the epidermis. Their biologically active forms can modulate the expression of genes involved in cellular differentiation and proliferation. They are expected to be helpful in 1) renewing epidermal cells, 2) acting as UV filters, 3) preventing oxidative stress, 4) controlling cutaneous bacterial flora, and 5) improving skin aging and photoaging. The genomic effects of topical retinoids most probably account for inducing many of the biologic effects involved in reversing and preventing extrinsic and intrinsic skin aging such as the prevention of MMP activation and oxidative stress and the regeneration of extracellular matrix. Inside the cells, they bind to nuclear receptors (RAR-α, -β, -γ, and RXR-α, -β, -γ) then the ligand-receptor complexes modulate the expression of genes involved in cellular differentiation and proliferation. In particular, retinoids are known to inhibit keratinocyte differentiation and to stimulate epidermal hyperplasia. Heparin binding-epidermal growth factor (HB-EGF) activation of keratinocyte ErbB receptors via a RAR-dependent paracrine loop has been proposed to mediate retinoid-induced epidermal hyperplasia [14].

How can we extend our knowledge of the interaction between cosmeceuticals and the genome? The development of nutritional genomics is an opportunity, which allows us to study genome-wide influences of cosmeceuticals/ nutriceuticals. A single nutrigenomics experiment can generate an enormous amount of data. Microarrays now make it possible to assess the effect of a specific diet or nutrient on the expression of a large proportion of the whole genome.

Nutritional Genomics and Cosmeceuticals/ Nutriceuticals

There is a growing realization that the effects of nutrition on health and disease cannot be understood without a profound understanding of how nutrients act at the molecular level. From a molecular standpoint, nutrients are considered to be "signaling molecules" that, through appropriate cellular sensing mechanisms, result in translation of these dietary signals into changes in gene, protein, and metabolite expression. The molecular structure of a nutrient determines the specific signaling pathways that it activates. Small changes in structure can have a profound influence on which sensor pathways are activated. The ω-3 polyunsaturated fatty acids have a positive effect on cardiac arrhythmia, whereas saturated C16-18 fatty acids do not.

The mechanisms of how some genes or transcription factors being activated by nutrients are now established. For example, the peroxisome proliferator activated receptor gamma 2 (PPAR-γ) is activated by the dietary lipids linoleic, linolenic, arachidonic, and eicosapentaenoic acid and their metabolites. Numerous receptors bind nutrients and their metabolites. These include retinoic acid (retinoic acid receptor (RAR) and retinoid X receptor (RXR)), fatty acids (peroxisome proliferator-activated receptors (PPARs) and liver X receptor (LXR)), vitamin D (vitamin D receptor (VDR)), oxysterols (LXR), bile salts (farnesoid X receptor (FXR)) or other hydrophobic food ingredients (CAR or PXR) [5].

In some situations the extracellular molecule itself enters the nucleus and binds a receptor; the resultant complex serves as the transcription factor (TF). In other situations, the TF can bind one or more additional molecules to form the TF complex, which then binds to the DNA. The binding of the TF causes a conformational change in the DNA and promotes or inhibits transcription, as needed for the appropriate response to the environment [7]. Bioactive dietary components are examples of this latter situation where these components do not themselves serve as TFs but influence the ability of TFs to bind to their DNA elements. For example, vitamin A modifies expression of several genes through the retinoic acid receptor TF. PUFAs influence the ability of peroxisome proliferator activator receptor/retinoic acid receptor complexes to bind to their DNA elements. Flavonoids regulate various genes through three TFs: nuclear factor- kappa B, estrogen receptor, and activating protein-1. In each case, the bioactive dietary components influence gene expression and thereby alter genetic outcomes [7]. DNA methylation may be altered through changes in nutrient intake since *S*-adenosylmethionine (*S*-AM), the substrate for DNA methyltransferases, is a product of the one carbon pathway fed with choline, methionine, and folate. Vitamin B-12, vitamin B-6, and riboflavin are cofactors for certain enzymes involved in these pathways. Deficiencies in these nutrients alter the carbon metabolism, impair DNA methylation, and have been linked to increased risk of neural tube defects, cancer, and CVDs [22].

Chromatin remodeling is dynamic on short time intervals and may also be altered permanently through changes in nutrient availability during key developmental windows. In addition to nutrient levels, total caloric intake has also been linked to chromatin remodeling [22, 23]. With the development of high throughput technologies, the following goals of nutrigenomics research can be defined: the identification of transcription factors that function as nutrient sensors and the genes they target; the elucidation of the signaling pathways involved, and characterization of the main dietary signals; the measurement and validation of cell- and organ-specific gene expression signatures of the metabolic consequences of specific micronutrients and macronutrients; the elucidation of the interactions between nutrient-related regulatory pathways and proinflammatory stress pathways, to understand the process of metabolic dysregulation that leads to diet-related diseases; the identification of genotypes that are risk-factors for the development of diet-related human diseases and quantification of their impact; and the use of nutritional systems biology to develop biomarkers of early metabolic dysregulation and susceptibility that are influenced by diet [5].

Dermagenetics (an off shoot of nutrigenetics) examines selected genetic mutations to determine what nutraceuticals will offer to benefit the individual skin health. Male pattern hair loss is an emotionally devastating condition. By the time hair loss or thinning is noticeable- almost 50% of hair could be gone. However, the only FDA approved medications to combat hair loss, minoxidil and finasteride, are most effective at stabilizing hair loss rather than re-growth. The earlier we can predict Male Pattern Loss, the more likely we are to save the hair. The HairDX Genetic Test for Male Pattern Hair Loss reports the presence or absence of a specific variation in the androgen receptor gene. Men who test positive for the variant AR gene have up to an 80% chance of developing hair loss. In contrast, men who test negative for the variant AR gene have up to a 90% chance of not developing hair loss. Men who test positive for the variant AR gene should be advised that they are at a high risk for developing Male Pattern Hair Loss. These men may wish to consult with a physician specializing in hair restoration, who can develop a plan to monitor and treat hair loss. For example, such men may be good candidates for minoxidil or finasteride (Propecia). Other developments in nutrigenetics, such as the AcneDX Genetic Test for antibiotic response of acne, also show us the good application prospect of personalized cosmeceuticals/ nutriceuticals.

Challenges of Nutritional Genomics and Conclusion

The cosmeceutical category is an undefined, unclassified, and unregulated area of dermatology that is yet in its infancy. Cosmeceuticals represent a category of products placed between nonprescription and prescription products [12]. The difference between a drug and a cosmeceutical is that the former is defined by having a biological effect on living tissue. Cosmeceutical products often are tested through in vitro studies using silicone replicas of skin and, at best, clinical trials are small, open-label studies usually supported by the cosmetic companies themselves [9]. The ability of these ingredients to enhance skin functioning depends on how they are formulated into products that can maintain the integrity of the active ingredient, deliver it in a biologically active form to the skin, reach the target site in sufficient quantity to exert an effect, and properly release from the carrier vehicle [19].

Cosmeceuticals may cause some unwanted problems, such as irritability to the skin, contact dermatitis, photosensitivity, comedogenicity, hair and nail damage, hyper- or hypopigmentation, infectivity, carcinogenicity, and even systemic adverse effects [10]. As such, careful preclinical, clinical and post-clinical evaluation of efficacy and safety is necessary for the development of a specific cosmeceutical product [10].

Moreover, we should not think that all 'nutrients' in foods have been characterized. There are probably thousands of chemicals present in foods, still classified as non-nutrients, which may play a significant role in gene regulation, and thus will significantly impact on health and disease [24]. The methodological challenges for genetic association studies and nutritional epidemiology can be addressed by firstly developing and using uniform nutrient intake measurements that can be compared within populations and among populations; secondly, standardizing measures of relevant physiological parameters; and thirdly, testing for genetic ancestry within chromosomal segments, rather than overall genetic make-up [22]. While these estimates are the best available for guiding preventive and treatment strategies, population-based risk fractions have limited utility for individuals: data derived from one ancestral group may not apply to individuals from other ancestral populations or to other individuals within the studied ancestral group because of gene-environment interactions [22].

The amount of micronutrients that appear to be protective against genome damage vary greatly between foods and careful choice is needed to design dietary patterns optimized for genome-health maintenance. As dietary choices vary between individuals, because of taste preferences that may be genetically determined or cultural or religious constraints, several options are required and supplements may be needed to cover gaps in micronutrient requirements. Clearly, the development of nutrient-dense foods and ingredients, such as aleurone flour, which is rich in bioavailable folate as well as other micronutrients required for DNA replication and repair, is essential in making it feasible for individuals to achieve their daily nutrient requirements for genome-health maintenance without intake of excess energy [21].

The recent explosion and fast-paced growth of the cosmeceutical market have left both consumers and dermatologists confused. Patients often receive unregulated information and efficacy claims for products from the internet and the media. Moreover, because many ingredients often are derived from plants, consumers may have a false sense of security that these "all-natural" products do not have potential adverse effects. In some instances, these products may not achieve the results they claim but they offer some benefit to the skin through an emollient effect. Indeed, it is often difficult to separate the effects of the moisturizer vehicle from the effects of the added active ingredient in cosmeceuticals [9]. The active ingredients in these natural products should be identified and evaluated using advanced technology to confirm efficacy and safety of their use in cosmeceutical products [12].

Some companies have utilized dermagenetics to lure customers into getting expensive genetic testing, and then recommend specific nutraceuticals or skin creams enriched by selected cosmeceuticals. However, there is no guarantee that a given mutation will indeed display significant abnormalities at the blood and tissue levels due to the contributions of post-translational modifications; and no evaluation of blood parameters are actually made in subjects before recommending a selected nutraceutical. The use of topical skin creams containing cosmeceuticals has developed into a multibillion dollar industry. How much benefit these cosmeceuticals actually provide when applied topically is currently a matter of debate. The degree of benefits achieved might depend on formulation methods, the degree to which the active compounds can maintain integrity and the duration of their effects, and factors affecting their release from the carrier vehicles [8]. Nutrigenomics raises ethical, legal and social issues particularly with respect to how the public may access nutrigenetic tests and associated nutritional and lifestyle advice. Current regulatory controversy focuses on potential harms associated with direct-to-consumer (DTC) marketing of nutrigenetic tests and especially the need to protect consumers from unreliable tests, false claims and unproven dietary supplements. Primary care physicians have minimal training in nutrition and genetics, and medical geneticists are in high demand and short supply. Dietetic practitioners are experts in nutrition science and interest in nutrigenomics is growing among members of this professional group. However, as with physicians, dietetics practitioners would require considerable training to bring nutrigenomics into their practice capacity [25].

It is essential that state regulatory agencies set clear guidelines for diagnostic laboratories, especially for those offering genetic tests directly to consumers; more strictly monitor health claims of ingredients in skin creams; and work with the universities to train physicians, nutritionists, and dietary professionals in nutrigenomics to meet the health challenges of this postgenomic era [8]. There must be excellent communication among scientists from different disciplines (eg, nutrition, molecular biology, medicine, genomics, bioinformatics), and there must be a high level of compatibility of the genomics data produced in the various laboratories.

References

[1] Ghosh, D. Personalised food: how personal is it? *Genes Nutr.* 2010; 5: 51-53.

[2] Ordovas, J. M., Mooser, V. Nutrigenomics and nutrigenetics. *Curr. Opin. Lipidol.* 2004; 15: 101-108.

[3] Afman, L., Müller, M. Nutrigenomics: from molecular nutrition to prevention of disease. *J. Am. Diet Assoc.* 2006; 106: 569-576.

[4] Subbiah, M. T. Nutrigenetics and nutraceuticals: the next wave riding on personalized medicine. *Transl. Res.* 2007; 149: 55-61.

[5] Müller, M., Kersten, S. Nutrigenomics: goals and strategies. *Nat. Rev. Genet.* 2003; 4: 315-322.

[6] Willett, W. C. Balancing life-style and genomics research for disease prevention. *Science* 2002; 296: 695-698.

[7] Debusk, R. M., Fogarty, C. P., Ordovas, J. M., Kornman, K. S. Nutritrional genomics in practice: where do we begin? *J. Am. Diet Assoc.* 2005; 105: 589-598.

[8] Subbiah, M. R. Application of nutrigenomics in skin health nutraceutical or cosmeceutical? *J. Clin. Aesthet. Dermatol.* 2010; 3: 44-46.

[9] Choi, C. M., Berson, D. S. Cosmeceuticals. *Semin. Cutan. Med. Surg*. 2006; 25: 163-168.

[10] Gao, X. H., Zhang, L., Wei, H. C., Chen, H. D. Efficacy and safety of innovative cosmeceuticals. *Clin. Dermatol*. 2008; 26: 367-374.

[11] Draelos, Z. D. Cosmeceuticals: undefined, unclassified, and unregulated. *Clin. Dermatol*. 2009; 27: 431-434.

[12] Amer, M., Maged, M. Cosmeceuticals versus pharmaceuticals. *Clin. Dermatol*. 2009; 27: 428-430.

[13] Yoshimura, K., Sato, K., Aiba-Kojima, E., et al. Repeated treatment protocols for melisma and acquired dermal melanocytosis. *Dermatol. Surg*. 2006; 32: 365-371.

[14] Sorg, O., Antille, C., Kaya, G., et al. Retinoids in cosmeceuticals. *Dermatol. Ther*. 2006; 19: 289-296.

[15] Farris, P. K. Topical vitamin C: a useful agent for treating photoaging and other dermatologic conditions. *Dermatol. Surg*. 2005; 31: 814-817.

[16] Beitner, H. Randomized, placebo-controlled, double blind study on the clinical efficacy of a cream containing 5% alpha- lipoic acid related to photoaging of facial skin. *Br. J. Dermatol*. 2003; 149: 841-849.

[17] Chiu, A. E., Chan, J. L., Kern, D. G., et al. Double-blinded, placebo-controlled trial of green tea extracts in the clinical and histologic appearance of photoaging skin. *Dermatol. Surg*. 2005; 31: 855-859.

[18] Lupo, M. P. Cosmeceutical peptides. *Dermatol. Surg*. 2005; 31: 832-836.

[19] Draelos, Z. D. Cosmeceuticals: undefined, unclassified, and unregulated. *Clin. Dermatol*. 2009; 27: 431-434.

[20] Ni, A., Mu, Y., Gulati, O. Treatment of melisma with pycnogenol. *Phytother. Res*. 2007; 16: 567-571.

[21] Bull, C., Fenech, M. Genome-health nutrigenomics and nutrigenetics: nutritional requirements or 'nutriomes' for chromosomal stability and telomere maintenance at the individual level. *Proc. Nutr. Soc*. 2008; 67: 146-156.

[22] Kaput, J. Nutrigenomics research for personalized nutrition and medicine. *Curr. Opin. Biotechnol*. 2008; 19: 110-120.

[23] Bishop, N. A., Guarente, L. Genetic links between diet and lifespan: shared mechanisms from yeast to humans. *Nat. Rev. Genet*. 2007; 8: 835-844.

[24] Ordovas, J. M., Mooser, V. Nutrigenomics and nutrigenetics. *Curr. Opin. Lipidol*. 2004; 15: 101-108.

[25] Castle, D., Ries, N. M. Ethical, legal and social issues in nutrigenomics: the challenges of regulating service delivery and building health professional capacity. *Mutat. Res*. 2007; 622: 138-143.

In: Nutrigenomics
Editor: M. T. Ravi Subbiah

ISBN: 978-1-62417-051-5

Chapter VIII

Development of Nutraceuticals and Cosmeceuticals as Epigenetic Modulators in Skin Health Care

Wim Vanden Berghe*
Laboratory of Eukaryotic Gene Expression and Signal Transduction (LEGEST), Department of Physiology, Ghent University, Ghent, Belgium
Laboratory of Protein Science, Proteomics and Epigenetic Signalling, Department of Biomedical Sciences, University Antwerp, Campus Drie Eiken, Wilrijk, Belgium

Abstract

In Western and traditional folk medicine, botanical polyphenols are widely used to treat skin diseases such as psoriasis and vitiligo, to prevent skin cancer, to accelerate skin wound healing, to reduce UV damage, to diminish skin inflammation or to preserve skin beauty. Skin diseases have a complex etiology and pathogenesis, including genetics and acquired epigenetic factors depending on the environment and diet. Aberrant epigenetic modifications (i.e., alterations of genomic DNA methylation patterns, of post-translational modifications of histones and of microRNA profiles) have recently been demonstrated in skin tumours and autoimmune-related skin disorders, by affecting key cellular pathways such as cell cycle regulation, cell signalling, differentiation, DNA repair, apoptosis, invasion and immune recognition. Accumulating data support efficacy of phytochemicals to modulate multiple biological processes at low concentrations through their activity as epigenetic modulators. The development of nutraceuticals or cosmeceuticals as epigenetic agents warrants further preclinical and clinical investigation to explore its beneficial effects in cancer treatment, (re)programming stem cell differentiation, wound healing, skin regeneration, tissue homeostasis, or in "correcting" epigenetic marks responsible for inflammatory skin diseases and ageing.

* Corresponding author: Wim Vanden Berghe. Ghent University, K.L.Ledeganckstraat 35, Ghent, Belgium. University Antwerp, Campus Drie Eiken, Universiteitsplein 1, Wilrijk, Belgium. E-mail: wim.vanden berghe@ua.ac.be.

Abbreviations

ATF3	activating transcription factor 3
Brg1	brahma-related gene 1
C/EBP	CCAAT enhancer binding protein
CARM	Coactivator-associated arginine methyltransferase
CREB	cAMP responsive element binding protein
DAMPs	danger associated molecular patterns
DNMT	DNA methyltransferase
ERK	extracellular regulated kinase
Ezh2	Enhancer of Zeste homolog 2
HAT	histone acetylase
HDAC	histone deacetylase
HMG	high mobility group
HS	hypersensitive site
hsp	heat shock protein
IκB	inhibitor of NFκB
IKK	IκB kinase
IRF3	interferon responsive factor 3
Jmjd3	jumonji domain containing 3 histone lysine demethylase
JNK	Jun N-terminal kinase
KDMT	lysine demethylase
KMT	lysine methyl transferase
MSK	Mitogen and stress activated *kinase*
NEMO	NFκB essential modulator
NIK	NFκB-inducing kinase
p/CAF	p300/CBP associated factor
PAMPs	pathogen associated molecular patterns
PARP	poly(ADP-ribose) polymerase
Pin 1	prolyl isomerase
PKA	protein kinase A
PML	promyelocytic leukemia
PRMT	protein arginine methyltransferase
SIRT1	sirtuin silent information regulators
SUMO	small ubiquitin-like modifier
SWI/SNF	SWItch/Sucrose NonFermentable nucleosome remodeling complex
TF	transcription factor

Keywords: epigenetic modulators; polyphenols; nutraceuticals; cosmeceuticals; DNA methylases; skin disorders; skin ageing; DNA methyl transferases; chromatin states; histone structure; chromatin states; skin immune-homeostasis; nuclear factor ϒB; reactive oxygen species

Introduction

Today, nutrition research has shifted from focusing exclusively on alleviating nutrient deficiencies to preventing diseases and composing manufactured functional foods (i.e. nutraceuticals and cosmeceuticals) with health protective properties. In an environment with constant unavoidable exposure to various toxicants, cells are constantly subjected to diverse noxious insults [1]. Free radicals and reactive oxygen species are synthesized endogenously (e.g., in energy metabolism and the antimicrobial defence system of the body) or produced as reactions to exogenous exposure (e.g. cigarette smoke, imbalanced diet, exhaustive exercise, environmental pollutants, food contaminants, carcinogens, dietary mutagens, microorganisms, solar radiation, etc.) [2-6]. Oxidative stress and inflammatory damages of cellular macromolecules play important roles in activating cytokines, oncogenes and destroying tumour suppressor genes, thereby setting up a ground for skin disorders and cellular transformation to tumours [7-9].

Skin, the biggest organ in mammals, protects the body from environmental hazards and prevents dehydration. Embryonic skin morphogenesis and homeostasis of adult skin require an accurately controlled gene expression in a spatiotemporally specific manner. Human dietary intervention studies based on the use of antioxidant compounds show how they can protect from endogenous and exogenous environmental assaults, and neutralize sun-induced effects on the skin [10]. Many plants are rich sources of bioavailable antioxidants. As such, humanity's original skin-care products were plant-based salves, ungents and poultices. Since oxidative stress is a smouldering issue in setting up an inflammation-cancer link at the early stage of carcinogenesis, reinforcing the body's antioxidant arsenal is the foremost strategy to cut the inflammation-cancer loop [11]. Chemoprevention, the strategy to inhibit, retard, or even reverse the specific stage of inflammation-carcinogenesis, is one of the most rational approaches to reduce the global burden of cancer [12]. One strategy is to combine topical and/or systemical applications of cosmeceuticals and nutraceuticals to prevent the damaging effects of ultraviolet radiation, inflammation, oxidative stress, noxious toxins and environmental pollutants on the many biologic processes involving epidermal homeostasis, premature skin ageing and cancer [13]. Phytochemicals have also succesfully been applied for regenerative medicine and cancer stem cell therapy [13-18]. Stem cells in the epidermis have a crucial role in maintaining tissue homeostasis by providing new cells to replace those that are constantly lost during tissue turnover of following injury [13, 19-21]. Different resident skin stem cell pools contribute to the maintenance and repair of the various epidermal skin tissues. Finally, plant polyphenols can also induce skin pathologies such as contact, allergic, and phytophotodermatitis [22].

Dietary agents consist of many bioactive ingredients which actively regulate various molecular targets. New evidence emerged that various dietary exposures may be vitally important in determining health beneficial effects, as a consequence of diet-induced epigenetic changes [23-26]. The diverse and broad biological activities of dietary phytochemicals are mediated through direct interaction with target proteins as well as epigenetic modulation of target genes, through its interaction with histone deacetylases, histone acetyltransferases, DNA methyltransferase I, and microRNA [27]. In this chapter, we will discuss how inflammation and various botanical phytochemicals can act as epigenetic modulators in inflammatory skin disorders, ageing or cancer [28-31]. The effects of chronic low-dose human exposure to dietary factors or environmental chemicals are not well characterized. Compounds or environmental stressors are routinely tested for the ability to induce DNA mutations, cause chromosome damage, or produce cell death, but are rarely tested for their ability to cause epigenetic changes, which can influence the behaviour of a cell without directly changing the DNA sequence [32-36]. Because epigenetic changes are reversible, developing drugs that control epigenetic regulation now attracts substantial research investment, including the development of functional foods or supplements as nutrition based epigenetic modulators for cancer chemoprevention [15, 37-45] In this respect, epigenetic changes have become the focus of intensive research in an attempt to understand the mechanisms by which they function in skin homeostasis and carcinogenesis [34, 46-48]. Interestingly the signalling mechanisms that control epithelial morphogenesis in the skin are reused during adult life to regulate skin homeostasis. As such, reversible programming of stem cell differentiation by phytochemicals with epigenetic modulator capacity has recently received much attention in cancer stem cell biology and regenerative applications [15-17, 49-52] (Figure 1).

Chromatin States in the Epigenomic Landscape

In general, DNA is wrapped around nucleosomes, which are arranged as regularly spaced beads (146 bp DNA/nucleosome) along the DNA. Typically, nucleosomes consist of a histone octamer of histones (H)2A/B, H3 and H4. The DNA bridging two adjacent nucleosomes is normally bound by the linker histone H1 and is termed linker DNA. While the core histones are bound relatively tightly to DNA, chromatin is largely maintained by the dynamic association with its architectural proteins.

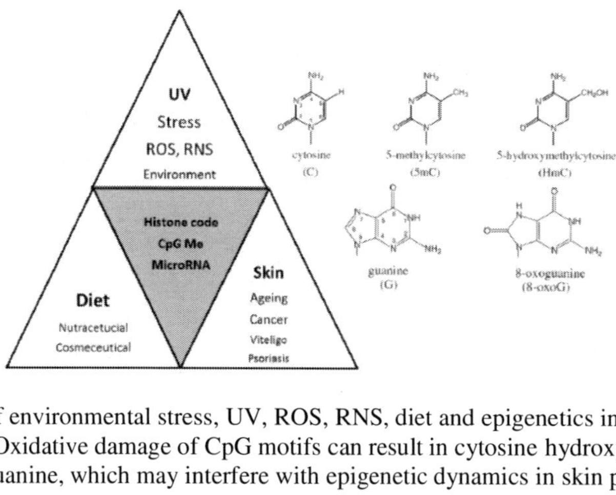

Figure 1. Interplay of environmental stress, UV, ROS, RNS, diet and epigenetics in skin homeostasis and carcinogenesis. Oxidative damage of CpG motifs can result in cytosine hydroxymethylation and formation of 8-oxoguanine, which may interfere with epigenetic dynamics in skin pathophysiology.

Before most activators of a gene access their DNA-binding sites, a transition from a condensed heterochromatin ("solenoid-like fiber") to a decondensed euchromatin ("beads on a string") structure appears to take place. Conversely, the acquisition of a more condensed heterochromatin structure is often associated with gene silencing [53]. This structural restriction of silenced chromatin on gene expression can be overcome by chromatin cofactor complexes, which remodel nucleosomes along the DNA or reversibly modify (acetylation, phosphorylation, ubiquitylation, glycosylation, sumoylation) histones on lysine, arginine, serine or threonine residues of amino-terminal histone tails. Since the discovery of histone modifying enzymes, N-terminal histone tails protruding from nucleosomes were found to be 'velcro patches' for polycomb proteins, (de)acetylases (HDAC/HAT), (de)methylases (HMT/HDMT), ubiquitin ligases, small *ubiquitin*-related modifier (SUMO) ligases, kinases, phosphatases, ribosylases, which together establish specific chromatin states involved in transcription [53-55]. Specific sets of histone modifications and/or variants are associated with genes that are actively transcribed or are repressed, a phenomenon defined as the "histone code" [53]. Based on coexisting histone marks and genome wide ChIP-seq data available within the ENCODE consortium, principal component analysis allowed to reduce the complexity of the histone code into 9 different chromatin states, linked to different functional regulatory features [54, 55].

To establish specificity of epigenetic marks, histone modifying complexes have to be recruited to relevant genomic locations by DNA-binding proteins, RNAs or protein-RNA complexes that bind to their specific DNA sites as a consequence of their own binding specificities and cellular concentrations [56-60]. It cannot come from the enzymatic activities per se as neither DNMTs, nor enzymes which modify histones know which part of the genome needs to be tagged. Furthermore, there is now a large body of evidence showing that modifications of the histone tails provide signals ("binary switches") that are recognized by specific binding proteins, such as chromo-, bromo- or tudor-domains which in turn can influence gene expression and other chromatin functions [61-63]. The time-dependent combinations of histone modifications or three-dimensional locus configuration further increase the complexity of information contained in chromatin [53, 62-67].

DNA methylation is the best-known epigenetic mark [68, 69]. It is catalyzed by two types of DNMTs: DNMT1 is a maintenance methyltransferase, whereas both DNMT3A and DNMT3B are *de novo* methyltransferases [70, 71]. The role of DNMT2 in DNA methylation is minor, its enzymology being largely directed to tRNA. DNA methylation is normally associated with gene inactivation and it usually occurs in CpG dinucleotides. Alternatively, DNA methylation of transcription factor binding sites which prevents binding of repressor proteins, may paradoxically induce gene activation. CpGs are normally methylated when scattered throughout the genome, but are mostly unmethylated when they are clustered as CpG islands at 5' ends of many genes. Hypermethylation of CpG-rich promoters triggers local histone code modifications resulting in a cellular camouflage mechanism that sequesters gene promoters away from transcription factors and results in stable silencing of gene expression. DNA methylation at CpG dinucleotides occurs upon transfer of S-adenosylmethionine (SAM) on cytosine by DNMTs. Whereas DNMT3A/B are responsible for DNA methylation during development (differentiation), DNMT1 is in charge of maintaining DNA methylation patterns in DNA replication during cell division. In mammalian cells, the fidelity of maintenance of methylation is 97–99.9% per mitosis, whereas de novo methylation is as high as 3–5% per mitosis, thus creating possibilities for epigenetic changes. DNA methylation also regulates genomic imprinting [72], X-chromosome inactivation [73] and silencing of repetitive sequences [74]. Although in most cases DNA methylation is a stable epigenetic mark, reduced levels of methylation can also be observed during development. This net loss of methylation can either occur passively by replication in the absence of funtional maintenance methylation pathways, or actively, by removing methylated cytosines. In plants active demethylation is achieved in plants by DNA glycosylase activity, probably in combination with the base excision repair pathway. In mammals, coupling of 5-methylcytosine deaminase and thymine DNA glycosylase activities may be responsible for DNA demethylation. Alternatively, a role for the 5-hydroxymethylcytosine modification in mammalian DNA demethylation has also been proposed as an intermediate in an active DNA demethylation pathway involving DNA repair and 5-hydroxymethylcytosine-specific DNA glycosylase activity [71]. Of particular interest, ROS and oxidative stress may affect DNA demethylation by DNA oxidation or Tet mediated hydroxymethylation [75, 76].

Although DNA methylation is the best-known epigenetic mark [70, 77, 78], DNA methylation does not act alone. It occurs in the context of nucleosome positioning and histone modifications [79-81]. For example, high resolution DNA methylation analysis has revealed 10-base periodicities (i.e one helical turn) in the DNA methylation status of nucleosome-bound DNA and found that nucleosomal DNA was more highly methylated than flanking DNA [81]. These data revealed that nucleosome positioning influences DNA methylation patterning of promoters and intron-exon boundaries throughout the genome and that DNA methyltransferases preferentially target nucleosome-bound DNA. Whether nucleosome strings provide a combinatory histone code is a matter of debate [53, 62, 80, 82-85], but in any event, histone modifications influence gene activity and regulation. For example, acetylation of lysines is generally associated with transcriptional activation whereas lysine methylation can dictate either activation (e.g. H3K4, H3K36, H3K79) or suppression (e.g. H3K9, H3K27 or H4K20). Specific histone modifications have been shown to be associated with DNA hypermethylation of CpG islands, including deacetylation of histones H3 and H4, loss of H3K4Me, and gain of H3K9Me3 and H3K27Me3 [86, 87]. DNA methylation marks are recognized by DNA methyl-binding proteins (MBD) which can interact with corepressor-associated enzymes (i.e. HDACs, enhancer of zeste homologue (EZH)2, ...), thus further linking DNA methylation and chromatin regulation [56, 88]. Altogether, "histone code" may only become biologically meaningful at the level of loop domains which, upon integration of conformations of multiple nucleosomes, translates allosteric changes into specific gene (cluster) activities, in order to establish specific regulatory programs at the genome level [53, 67, 89-91]. In analogy to allosteric control of enzymes, specific gene activity may be determined by the spatial organization (compartmentalization in discrete territories) and structural landscape (three-dimensional structure) of a gene locus, by altering the higher order structure of chromatin (cis mechanism) or by generating a binding platform for effector proteins (trans mechanisms) [64, 65, 67, 89, 90, 92, 93].

There is good evidence that also noncoding RNAs regulate chromatin architecture [60, 83, 94-98]. The term non-coding RNA (ncRNA) is commonly employed for RNA that does not encode a protein. Although it has been generally assumed that most genetic information is transacted by proteins, recent evidence suggests that the majority of the genomes of mammals and other complex organisms is in fact transcribed into ncRNAs, many of which are alternatively spliced and/or processed into smaller products. Besides tRNA and rRNA, these ncRNAs include long-noncodingRNAs (lncRNAs), microRNAs (miRNAs) and tinyRNAs (tiRNAs) as well as several other classes of, sometimes yet-to-be-discovered, small regulatory RNAs such as snoRNAs [60, 94-96]. These RNAs (including those derived from introns) appear to comprise a hidden layer of internal signals that control various levels of gene expression in physiology and development, including chromatin architecture/epigenetic memory, transcription (enhancer function), RNA splicing, editing, translation and turnover [99]. RNA regulatory networks may determine most of our complex characteristics and play a significant role in disease [99]. For example, miRNAs can change expression levels of the epigenetic machinery (DNMT, HDAC, sirtuin (SIRT), polycomb (Pc) proteins, etc.) by posttranscriptional gene regulation involving base pairing with 3'untranslated (UTR) regions in their target mRNAs resulting in mRNA degradation or inhibition of translation [40, 83, 100-104]. Alternatively, long ncRNAs and tiRNAs can regulate gene expression and/or DNA methylation by promoter association [60, 94, 95, 98, 99]. DNA-methylation can thus also be RNA-directed [83, 105].

Controlling Immune Homeostasis in Skin Biology

Inflammation is a fundamental adaptation to the loss of cellular and tissue homeostasis covering many important processes, including host defence, tissue remodelling and repair, and the regulation of metabolism. The complexity of the inflammatory response requires that its many functional programs are controlled coordinately in some situations but independently in others [21, 106]. This is achieved through multiple mechanisms that operate at different levels, including alterations in the composition of immune cells in tissues, changes in cell responsiveness to inflammatory stimuli, regulation of signalling pathways and epigenetic control of inflammatory gene expression. First, cell-specific mechanisms operate at the level of different cell types, and include regulation of their recruitment and activation. The skin is a complex and highly specialized organ with many different cell types which protect the body from water loss and adverse environmental stress, but also has important immunological properties, involving cell specific responses to its microenvironment. The skin is composed of two main compartments, the epidermis and the dermis, which communicate extensively to establish, maintain and restore homeostasis. The uppermost skin layer bordering environment is stratum corneum, a unique, highly lipophilic two-compartment system of enucleated cells. The underlying layer (epidermis) contains keratinocytes at different stages of differentiation. The underlying dermal layer is rich of various types of cells, namely fibroblasts, melanocytes, endothelial cells, smooth muscle cells, immune cells (lymphocytes, granulocytes, monocytes, and dendritic cells, Langerhans cells, mast cells), and neurons. Disturbance of skin immune homeostasis causes chronic inflammatory diseases such as psoriasis. In inflamed tissues, a wide variety of activated inflammatory-immune cells (e.g. mast cells, neutrophils, leukocytes, macrophages, monocytes, eosinophils, dendritic cells, phagocytes and natural killer cells) generate reactive oxygen species (ROS) and/or reactive nitrogen species (RNS) which can inactivate DNA repair enzymes and cause DNA damage, chromatin modifications, epigenetic changes, variation in microRNA profiles or genome instability, all promoting carcinogenesis [12, 107, 108].

Second, signal-specific mechanisms operate at the level of pathways which for example activate the key transcription factor nuclear factor-κB (NFκB). Members of the nuclear factor κB (NFκB) family of dimeric transcription factors (TFs) regulate expression of a large number of genes involved in immune responses, inflammation, cell survival, and cancer [109, 110]. NFκB TFs are rapidly activated in response to various stimuli, including cytokines, UV, infectious agents or danger signals (bacteria, viruses, chemicals, pathogen associated molecular patterns (PAMPs), danger associated molecular patterns (DAMPs), and radiation-induced DNA double-strand breaks [2-6, 111, 112]. In nonstimulated cells, some NFκB TFs are bound to inhibitory IκB proteins and are thereby sequestered in the cytoplasm. NFκB activation by IκB kinase leads to phosphorylation of IκB proteins and their subsequent recognition by ubiquitinating enzymes. The resulting proteasomal degradation of IκB proteins liberates NFκB TFs, which translocate to the nucleus to drive expression of target genes. IκB kinase alpha (IKKα), IKKβ, and IKKγ/NEMO form the IKK complex, which is essential for NFκB activation and histone H3 modifications. Loss of IKKα has been shown to cause epidermal hyperplasia, prevent keratinocyte terminal differentiation, and impair the formation of the skin, resulting in the deaths of IKKα-deficient (IKKα-/-) mice soon after birth. Recent experimental data from several laboratories have revealed that IKKα functions as a tumor suppressor in human squamous cell carcinomas (SCCs) of skin, lungs, and head and neck. Chemical carcinogenesis studies using mice have shown that reduction in IKKα expression increases the number and size of Ras-initiated skin tumors and promotes their progression, indicating that reduced IKKα expression provides a selective growth advantage that cooperates with Ras activity to promote skin carcinogenesis [113-115]. Alternative to IKK, various other kinases have been identified which modulate transcriptional nuclear activity of NFκB and chromatin dynamics, including MSK, PKAc, PI3K, Akt [113, 116-119]. Besides NFκB, growth factors, hormones and a variety of environmental stresses also enhance Jun/AP1 activity, e.g. through phosphorylation thereby regulating cell proliferation, differentiation, transformation and/or apoptosis [120, 121]. The Jun proteins Jun, JunB and JunD are core members of activator protein-1 (AP-1), a dimeric transcription factor complex consisting of homo- and heterodimers of the Jun, Fos, activating transcription factor (ATF) and musculoaponeurotic fibrosarcoma (Maf) families. Regulation of AP1 can be achieved at different levels by changes in transcription of genes encoding AP1 subunits, by controlling the stability of the mRNAs, by post-translational processing, turnover and modification by phosphorylation of AP1 proteins, and by specific interaction between AP1 proteins and other transcription factors and cofactors. The mechanism of post-translational control is extensively documented in the case of mitogen- and cellular-stress-induced hyperphosphorylation of Jun through the Jun N-terminal kinase (JNK) cascade. Additional kinases such as casein kinase II, glycogen synthase kinase 3β or RSK2 can phosphorylate Fos and Jun proteins thereby regulating their transactivation potential and DNA-binding activity. High levels of phosphorylated c-Jun, Fra1, Fra2 and ATF-2 proteins correlate with invasive gene expression programs and malignant phenotypes in the multistage mouse skin carcinogenesis model [122-125]. Tumours are typically infiltrated with immune cells and inflammation impacts most, if not all, stages of tumorigenesis [7-9, 126-129]. Tumours have often been referred to as 'wounds that never heal' and regularly involve hyperactivated NFκB and AP1 members. Processes involved in wound healing, such as activation of matrix remodelling, cell motility and angiogenesis, are also common features of tumour progression [130]. Although the cellular and molecular mechanisms by which NFκB/AP1 members disturb skin immune homeostasis and trigger skin inflammation in the epidermis remain elusive, it seems that impaired NFκB/AP1 signalling in keratinocytes elicits a response that has many similarities to the wound-healing reaction [21]. It is probable that the NFκB/AP-1-deficient epidermis releases stress signals that are interpreted by stromal and immune cells in the dermis as an injury, eliciting a wound-healing response. The skin is routinely exposed to various stresses that could challenge NFκB/AP1 activity in the epidermis [21, 124, 131]. For example, UV damage, environmental antigens, temperature changes, mechanical stress or even exposure to the air could provide such signals. Ultraviolet (UV) radiation is one of the most abundant carcinogens in our environment which induces the formation of reactive oxygen species (ROS) and triggers the development of non-melanoma skin cancers. ROS molecules damage cellular macromolecules, resulting in premature aging, severe tissue destruction or even carcinogenesis. Large amounts of ROS are also generated by inflammatory cells in skin wounds as a defence against bacterial infection. Although this activity is beneficial, ROS can damage the inflammatory cells themselves and other cells at the wound site. Therefore, cells had to develop strategies to detoxify ROS, such as production of low molecular weight antioxidants and expression of ROS-detoxifying enzymes. The Nrf2 transcription factor plays a key role in the cellular defence against UV, oxidative and xenobiotic stresses through its capability to induce the expression of genes, which encode detoxifying enzymes and antioxidant proteins [132, 133]. Nrf2 is normally retained in the cytoplasm via interaction with Keap1, an actin-binding protein, which also mediates its degradation via the ubiquitin–proteasome pathway. Following electrophilic activation of Keap1, Nrf2 gets stabilized. This results in accumulation in the nucleus, where it binds to cis-acting antioxidant response elements (ARE) in the promoters of its target genes, for example, NADPH-quinone oxidoreductase (NQO1), several glutathione S-transferases (GST), -glutamyl-cysteine synthetase heavy subunit (-GCSh) and light subunit (-GCSl), and heme oxygenase 1 (HO-1). Many phytochemicals that belong to various families of secondary metabolites, such as alkaloids (caffeine, sanguinarine), flavonoids [(-)-epigallocatechin 3-gallate, genistein, silibinin], carotenoids (beta-carotene, lycopene), and isothiocyanates (sulforaphane), hold promise for the development of protective strategies [134]. Mechanistically, they affect multiple signalling pathways and protect against UV radiation-inflicted damage by their ability to act as direct and indirect antioxidants, as well as anti-inflammatory and immunomodulatory agents. Such "pluripotent character" is a critical prerequisite for an agent that is designed to counteract the multiple damaging effects of UV radiation (Figure 2).

Thirdly, gene-specific mechanisms operate at the level of individual genes and gene clusters. Induction of inflammatory transcriptional responses is orchestrated by many transcription factors, consistent with the complexity of the response. Controlled expression of cytokine genes is an essential component of an immune response and is crucial for homeostasis. In order to generate an appropriate response to metabolic stress or an infectious condition, the type of cytokine, as well as the cell type, dose range and the kinetics of its expression are of critical importance [135-137]. Selectivity is conferred by the expression of specific NFκB subunits and their respective posttranslational modifications, and by combinatorial interactions between NFκB and other transcription factors, such as activator protein (AP-1), signal transducers and activators of transcription (STAT3), nuclear factor erythroid 2-related factor (Nrf2), peroxisome proliferator-activated receptor- γ (PPAR γ), estrogen receptor (ER), liver X receptor (LXR), hypoxia inducible factor-1 (HIF-1) and p53 which are involved to a variable extent in immune homeostasis, angiogenesis, chemoresistance, stem cell survival, cancer invasion and tumour progression [9, 124, 138-141]. Various naturally occurring phytochemicals such as withaferin, curcumin, resveratrol, mangiferin hold promise as anti-cancer drugs by interfering with NFκB, STAT3, AP1, HIF, PPAR, ER, LXR, p53 activities and gene expression programs [142-146]

From Enhanceosome to Histone Code

Since transcription factors (TF) bind very poorly or not at all to nucleosomal DNA, their activation is coordinated to recruitment of noncoding RNAs [60, 98], DNMTs [59] and epigenetic writer, reader or eraser proteins [53], including ATP-dependent chromatin-remodelling factors (swith/sucrose non fermentable (SWI/SNF), Brahma (Brm), brahma-related gene, (Brg1)), histone-enzyme complexes such as kinases (IKK, MSK, ataxia telangiectasia mutated (ATM), AKT, PI3K), poly(ADP-ribose) polymerase (PARP), methylases (EZH2, coactivator-associated arginine methyltransferase (CARM)1, protein arginine methyltransferases (PRMT)), demethylases (lysine specific demethylase (LSD)1, Jumonji C family histone demethylase (JMJD)3), prolyl isomerase (PIN1), acetylases (p300, CREB binding protein (CBP), p300/CBP associated factor (p/CAF)), deacetylases (HDAC, SIRT) and DNMTs [56, 116, 147-150].

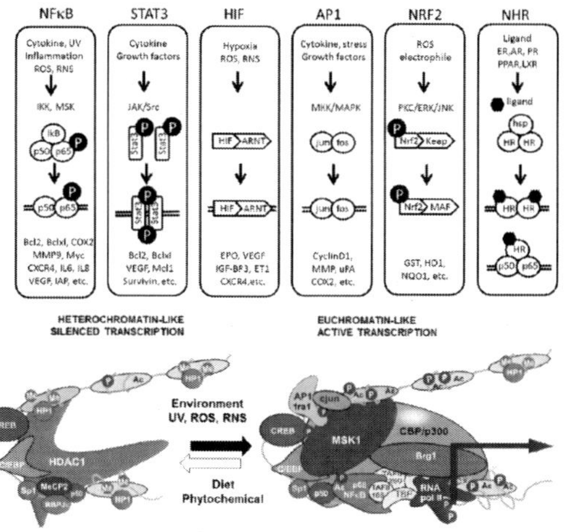

Figure 2. Dietary modulation of transcription factor pathways which regulate chromatin "oscillation" dynamics between euchromatic and heterochromatic gene regions.

Parallel posttranslational modifications (phosphorylation, acetylation, methylation, ribosylation, sumoylation, ubiquitination) of histone and non-histone TF and cofactor complexes in response to signalling components allow displacement of polycomb complexes and formation of dynamic enhanceosome complexes which establish a distinct chromatin structure [63, 116, 151-156]. These epigenetic settings are the ultimate integration sites of both environmental and differentiative inputs, determining proper expression of each target gene [110, 148, 150, 157, 158].

Investigation of gene expression and epigenetic regulation of inflammatory genes, revealed different subclasses according to chromatin activation mode, nucleosome stability and presence of CpG islands in the promoter [159, 160]. Furthermore, variable CpG DNA promoter methylation of NFκB pathway components IKK, IκB and RelB [161-163] was found to rewire gene induction properties upon exposure to a secondary inflammatory stimulus [161, 164-167].

Dietary Phytochemicals As Multi-Functional Epigenetic Modulators

Epigenetic changes arise in adults either by random change or upon exposure to environmental factors, such as stress, viral/bacterial infections, UV exposure, hormones toxicants and endocrine disruptors [150, 168, 169]. The demonstration that also diet can indirectly modulate gene expression by heritable changes in chromatin dynamics and DNA function through F1 and F2 generations launched much interest in nutritional epigenetics [38, 170-174]. More particularly, mice fed with a soy polyphenol diet reveal epigenetic changes in DNA methylation patterns in their offspring which protect against diabetes, obesity and cancer [38, 175-177]. Although effects of dietary factors and extracts have frequently been demonstrated in *in vitro* experiments at concentrations which can never be achieved *in vivo*, "epigenetics" sheds a more realistic light on dietary studies, as longlife exposure at physiological concentrations can remodel the epigenome in a cumulative fashion by repetitive effects on the epigenetic machinery [178-182]. Furthermore, It should be evaluated which epigenetic changes are stable over time.

Today, nutritional compounds (epicatechin, resveratrol, genistein, curcumin, isothiocyanates ...) have been characterized which interfere with enzymatic activity of DNA methyltransferase (DNMT), histone deacetylases (HDAC) and sirtuins (SIRT) which modulate inflammatory responses and immunological senescence [38, 43, 183-186]. Polyphenol compounds can change DNA methylation patterns either by blocking DNMT activity, DNA oxidation or else by interfering with methyldonor or acetyldonor metabolism [187, 188]. Interestingly, transient exposure to a specific diet can induce long-lasting activating epigenetic changes in the promoter of the nuclear factor κB (NFκB) subunit p65, which affects inflammatory gene expression [167]. Alternatively, compounds may chemically interfere with histone mark interacting effector domains (such as chromo-, bromo- or tudor domains) [61, 189, 190]. From another perspective, chemopreventive phytochemicals may indirectly modulate chromatin dynamics and epigenetic effects upon interference with global cell metabolism [191] (Figure 3).

NAD, acetyl-coenzyme A (Acetyl-coA) and S-adenosylmethionine (SAM) are elemental for epigenetic control of transcription including methylation of DNA and posttranslational modifications of histones and non-histone chromatin factors (not shown). NAD contributes to transcriptional control mainly via the activity of the protein deacetylase sirtuin (SIRT), which uses NAD as one of the substrates.

Sirtuins are also important for maintaining the activity of the acetyl-coA acetyltransferases. Ac-coA is synthesized by acetyl-coA-synthetase and ATP-citrate lyase that use acetate and citrate as the precursor, respectively. Citrate is an intermediate/product of the TCA cycle. SAM is the methyl donor for DNA, RNA, histones and non-histone protein methylation. SAH generated in each round of methylation reaction is a potent inhibitor of methyltransferases and has to be cleared by SAH hydrolase. NAD is an essential coenzyme for SAHH. Synthesis of methionine from homocystein is achieved through extracting the methyl group from betaine, derived from choline, or 5-methyl-THF, a derivative of folic acid. Metabolism of phospholipids and folic acid may thus indirectly contribute to epigenetic regulation. Likewise, the abundance of NAD and citrate is linked to the cellular energy flux, e.g. the TCA cycle. Changes in the expression of certain genes may therefore be influenced significantly. Abbreviations used: Ac-coA, acetyl-coenzyme A; ACS, acetyl-coA-synthetase; AC-ACS acetylated-ACS; Ado, adenosine; HAT, histone acetyltransferase; Hcy homocysteine; MTases, methyltransferases; NAD, Nicotinamide adenine dinucleotide; ROS, reactive oxygen species, RNS, reactive nitrogen species, SAH, S-adenosyl homocysteine; TCA tricarboxylic cycle; THF tetrahydrofolate.

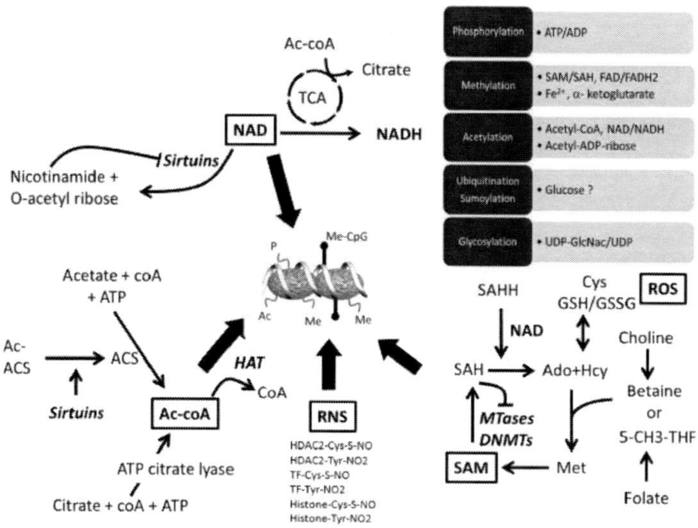

Figure 3. Once critical facet of histone modifications is that they are elicited by specific enzymatic activities that depend on the intracellular levels of essential metabolites: these metabolites sense ROS, RNS, cellular metabolism, nutrients and energy levels in the cell.

As such, epigenetic changes may follow biochemical metabolisation of dietary factors, which can deplete cellular pools of acetyl-CoA, NAD^+ and methyldonors, resulting in unbalanced DNA methylation and/or protein acetylation or methylation [184, 187, 192]. For example, flavanol-rich diets interfere with the methyldonor metabolism and the available pool of S-adenosylmethionine, resulting in changes in DNA methylation [193-197] and histone methylation, which is also affected by alterations in SAM/SAH ratios [198, 199]. Furthermore, even without nutritional deficiency of methyl groups, impaired synthesis of SAM and perturbed DNA methylation can happen when the need for the synthesis of the detoxification enzyme glutathione transferase (GSH) synthesis increases [200]. Diets or nutritional compounds which affect energy metabolism or mitochondrial respiration can have global epigenetic effects upon changes in NAD^+ availability and SIRT activity [201]. Along the same line, flavanol-rich diets interfere with the methyldonor metabolism and the available pool of S-adenosylmethionine, resulting in changes in DNA and histone methylation [194-197] [198, 199]. As such, specific dietary classes of functional food maybe designed as therapeutic epigenetic modulators in skin homeostasis.

Cell longevity and ageing is critically linked to activity of NAD-dependent deacetylase activity of the sirtuin family (SIRTs), which coordinates cell functions to optimize cells struggle to survive in a stressful environment, as is the case for skin cells [201, 202]. In this respect, sirtuin-target pharmaceutical, nutriceutical and cosmeceutical interventions against skin-ageing are also on the horizon [203-205]. Various plant polyphenols have already been identified as sirtuin-activating compounds, including butein, piceatanool, fisetin, quercetin and resveratrol [186]. Furthermore, a number of compounds have been further optimized as multifunctional ligands which can simultaneously inhibit KMT, HAT, SIRT and PRMT [206]. Since epigenetic changes are reversible, much attention is currently focusing on how modulation of hyper/hypomethylation status of key inflammatory genes by dietary factors can be applied as an effective approach to protect against inflammatory skin disease, cancer or ageing [26, 171, 174, 183]. As such, it will be a challenge for future anti-inflammatory therapeutics and dermatological research to identify novel epigenetic targets which allow selective modulation of the inflammatory signaling network [144, 207-212].

Inflammatory Skin Disorders and Epigenetic Reprogramming

Recent studies indicate that aberrant DNA methylation is a common feature of many human disorders including skin tumors (i.e. cutaneous carcinoma, melanoma, lymphoma) and auto-immune-related skin disorders (i.e. psoriasis, atopic dermatitis, vitiligo, lupus erythematosus). Skin cancer is driven by oncogene activation and the inactivation of tumor suppressor genes (TSGs). Although there is a wide range of established tumor models involving genetic lesions, the timing and hierarchy of global or specific epigenetic changes in tumor progression and metastasis are still poorly understood [213]. Briefly, the cancer epigenetic mechanisms can be summed up by the following three points: First, promoter hypermethylation of TSGs leads to gene inactivation. Second, hypomethylation of the proto-oncogene sequences results in the increased expression of proto-oncogenes and genomic instability. Third, DNA methylation increases the rate of mutation, affecting the genetic mechanisms underlying tumor development. Recent studies have demonstrated that the regulatory sequences of many TSGs are hypermethylated in various types of skin diseases and cancers, including proteins involved in cell adhesion, cell cycle, apoptosis, signaling, DNA repair, metabolism (for overview, see [214-217]). In recent years, several studies have demonstrated that disruption of epigenetic mechanisms can alter immune function and aberrant DNA methylation contributes to the development of autoimmune-related skin diseases such as lupus, psoriasis, atopic dermatitis, and vitiligo [215].

Besides skin disorders, repair of a defect in the human skin or wound healing is a highly orchestrated epigenetic process of inflammation, proliferation and maturation in order to re-establish barrier function by regenerating new skin. For example, polycomb-mediated silencing is lost during wound healing to enable upregulation of repair genes and demethylases Jmjd3 and Utx which control epidermal differentiation [131, 218, 219]. The inducible expression and repression of genes represents a key component of this regenerative process. Furthermore, it is critically important to recognize that the understanding of cutaneous wound healing is incomplete without appreciating the functional significance of wound-induced miRNA which play a key role in skin morphogenesis and in regulating angiogenesis. Furthermore, deregulation of specific miRNA is critical in derailing the healing sequence in chronic problem wounds [220-225].

Conclusion

In summary, it is increasingly clear that epigenetics is a crucial component of phenotypic variation in skin health care and disease. Better understanding of the role of the environment's interactions with the genome, for example UVR, skin ageing and cancer, or infections and autoimmune skin disease should lead to improvements in prevention of cutaneous disorders. Additionally it is likely that understanding and manipulating the epigenome, a potentially reversible source of biological variation, has great potential in the chemoprevention and treatment of skin disorders or ageing. Future research should focus on exploring epigenetic methods for the early diagnosis and monitoring of skin cancer. Also unlike genetic lesions, development of novel therapies using dietary/botanical inhibitors of DNMT, HDAC and K(D)MT have the potential to reverse epigenetic changes in skin disease. For example, chemopreventive effects of epigallocatechins in skin cancer could be attributed regulation of the Bmi1 polycomb group gene involved in epigenetic gene silencing [226]. Furthermore, for dihydrocoumarin (DHC), a compound found in Melilotus officinalis (sweet clover) that is commonly added to food and cosmetics, disrupted heterochromatic silencing as well as inhibition of human SIRT1 deacetylase activity could be demonstrated [227]. In another study, diet composition exacerbates inflammation and cell survival signals in the skin following UV irradiation [228]. However, there is some concern that epigenetic therapy with 'nonselective' inhibitors of DNMT, HDAC, HMT at longterm may suffer from lack of specificity [189, 206, 229]. As such, one may combine nonselective epigenetic therapies with more targeted approaches [59]. For example, combination treatment of specific transcription factor inhibitors and/or selective hormone receptor ligands with epigenetic drugs may trigger synergistic activities at subsets of inflammatory genes [230-237]. Finally, microRNA therapeutics also hold promise to join soon the arsenal of epigenetic drugs, as their target sequence specificity may bridge the gap between genetic and epigenetic changes [83, 220-223, 238]. The precise epigenetic effects of topical or systemical clinical applications of dietary botanicals in inflammatory skin disorders, skin ageing or skin regeneration, however, require further exploration.

Acknowledgment

This research is partially financial supported by an NOI grant (UA).

References

[1] Korkina, L. G., C. De Luca, V. A. Kostyuk, S. Pastore, Plant polyphenols and tumors: from mechanisms to therapies, prevention, and protection against toxicity of anti-cancer treatments. *Curr. Med. Chem.*, 2009; 16: 3943-3965.

[2] Carta, S., P. Castellani, L. Delfino, S. Tassi, R. Vene, A. Rubartelli, DAMPs and inflammatory processes: the role of redox in the different outcomes. *J. Leukoc. Biol.*, 2009; 86: 549-555.

[3] Srikrishna, G., H. H. Freeze, Endogenous damage-associated molecular pattern molecules at the crossroads of inflammation and cancer. *Neoplasia*, 2009; 11: 615-628.

[4] Rakoff-Nahoum, S., R. Medzhitov, Toll-like receptors and cancer. *Nat. Rev. Cancer,* 2009; 9: 57-63.

[5] Rakoff-Nahoum, S., R. Medzhitov, Role of toll-like receptors in tissue repair and tumorigenesis. *Biochemistry* (Mosc.), 2008; 73: 555-561.

[6] De Diego, J. L., G. Gerold, A. Zychlinsky, Sensing, presenting, and regulating PAMPs. *Ernst Schering Found. Symp. Proc*., 2007: 83-95.

[7] Grivennikov, S. I., M. Karin, Inflammation and oncogenesis: a vicious connection. *Curr. Opin. Genet. Dev*.; 20: 65-71.

[8] Aggarwal, B. B., Inflammation, a silent killer in cancer is not so silent! *Curr. Opin. Pharmacol*., 2009; 9: 347-350.

[9] Aggarwal, B. B., P. Gehlot, Inflammation and cancer: how friendly is the relationship for cancer patients? *Curr. Opin. Pharmacol*., 2009; 9: 351-369.

[10] Katiyar, S. K., Grape seed proanthocyanidines and skin cancer prevention: inhibition of oxidative stress and protection of immune system. *Mol. Nutr. Food Res*., 2008; 52 Suppl. 1: S71-76.

[11] Aggarwal, B. B., M. E. Van Kuiken, L. H. Iyer, K. B. Harikumar, B. Sung, Molecular targets of nutraceuticals derived from dietary spices: potential role in suppression of inflammation and tumorigenesis. *Exp. Biol. Med.* (Maywood), 2009; 234: 825-849.

[12] Kundu, J. K., Y. J. Surh, Nrf2-Keap1 Signaling as a Potential Target for Chemoprevention of Inflammation-Associated Carcinogenesis. *Pharm. Res*., 2010; in press.

[13] Blanpain, C., E. Fuchs, Epidermal homeostasis: a balancing act of stem cells in the skin. *Nat. Rev. Mol. Cell. Biol*., 2009; 10: 207-217.

[14] Crea, F., L. A. Mathews, W. L. Farrar, E. M. Hurt, Targeting prostate cancer stem cells. *Anticancer Agents Med. Chem*., 2009; 9: 1105-1113.

[15] Kawasaki, B. T., E. M. Hurt, T. Mistree, W. L. Farrar, Targeting cancer stem cells with phytochemicals. *Mol. Interv*., 2008; 8: 174-184.

[16] Shytle, R. D., J. Ehrhart, J. Tan, J. Vila, M. Cole, C. D. Sanberg, P. R. Sanberg, P. C. Bickford, Oxidative stress of neural, hematopoietic, and stem cells: protection by natural compounds. *Rejuvenation Res*., 2007; 10: 173-178.

[17] Bickford, P. C., J. Tan, R. D. Shytle, C. D. Sanberg, N. El-Badri, P. R. Sanberg, Nutraceuticals synergistically promote proliferation of human stem cells. *Stem. Cells Dev*., 2006; 15: 118-123.

[18] Zhou, J., H. Zhang, P. Gu, J. Bai, J. B. Margolick, Y. Zhang, NF-kappaB pathway inhibitors preferentially inhibit breast cancer stem-like cells. *Breast Cancer Res. Treat*., 2008; 111: 419-427.

[19] Oplander, C., C. V. Suschek, New aspects of nitrite homeostasis in human skin. *J. Invest. Dermatol*., 2009; 129: 820-822.

[20] Zaph, C., A. E. Troy, B. C. Taylor, L. D. Berman-Booty, K. J. Guild, Y. Du, E. A. Yost, A. D. Gruber, M. J. May, F. R. Greten, L. Eckmann, M. Karin, D. Artis, Epithelial-cell-intrinsic IKK-beta expression regulates intestinal immune homeostasis. *Nature*, 2007; 446: 552-556.

[21] Pasparakis, M., Regulation of tissue homeostasis by NF-kappaB signalling: implications for inflammatory diseases. *Nat. Rev. Immunol*., 2009; 9: 778-788.

[22] Korkina, L. G., S. Pastore, C. De Luca, V. A. Kostyuk, Metabolism of plant polyphenols in the skin: beneficial versus deleterious effects. *Curr. Drug Metab.*, 2008; 9: 710-729.

[23] Diet, Epigenetic Events, and Cancer Prevention. Proceedings of a symposium, September 26-27, 2007. *Nutr. Rev.*, 2008; 66 Suppl. 1: S1-72.

[24] De Assis, S., L. Hilakivi-Clarke, Timing of dietary estrogenic exposures and breast cancer risk. *Ann. N. Y. Acad. Sci.*, 2006; 1089: 14-35.

[25] Guerrero-Bosagna, C. M., P. Sabat, F. S. Valdovinos, L. E. Valladares, S. J. Clark, Epigenetic and phenotypic changes result from a continuous pre and post natal dietary exposure to phytoestrogens in an experimental population of mice. *BMC Physiol.*, 2008; 8: 17.

[26] Hunter, P., We are what we eat. The link between diet, evolution and non-genetic inheritance. *EMBO Rep.*, 2008; 9: 413-415.

[27] Li, Y., T. O. Tollefsbol, Impact on DNA methylation in cancer prevention and therapy by bioactive dietary components. *Curr. Med. Chem.*, 2010; 17: 2141-2151.

[28] Fraga, M. F., R. Agrelo, M. Esteller, Cross-talk between aging and cancer: the epigenetic language. *Ann. N. Y. Acad. Sci.*, 2007; 1100: 60-74.

[29] Fraga, M. F., E. Ballestar, M. F. Paz, S. Ropero, F. Setien, M. L. Ballestar, D. Heine-Suner, J. C. Cigudosa, M. Urioste, J. Benitez, M. Boix-Chornet, A. Sanchez-Aguilera, C. Ling, E. Carlsson, P. Poulsen, A. Vaag, Z. Stephan, T. D. Spector, Y. Z. Wu, C. Plass, M. Esteller, Epigenetic differences arise during the lifetime of monozygotic twins. *Proc. Natl. Acad. Sci. US*, 2005; 102: 10604-10609.

[30] Fraga, M. F., M. Esteller, Epigenetics and aging: the targets and the marks. *Trends Genet.*, 2007; 23: 413-418.

[31] Kahn, A., M. F. Fraga, Epigenetics and aging: status, challenges, and needs for the future. *J. Gerontol. A Biol. Sci. Med. Sci.*, 2009; 64: 195-198.

[32] Ikehata, H., H. Kudo, T. Masuda, T. Ono, UVA induces C-->T transitions at methyl-CpG-associated dipyrimidine sites in mouse skin epidermis more frequently than UVB. *Mutagenesis*, 2003; 18: 511-519.

[33] Ikehata, H., T. Masuda, H. Sakata, T. Ono, Analysis of mutation spectra in UVB-exposed mouse skin epidermis and dermis: frequent occurrence of C-->T transition at methylated CpG-associated dipyrimidine sites. *Environ. Mol. Mutagen.*, 2003; 41: 280-292.

[34] Ikehata, H., T. Ono, Significance of CpG methylation for solar UV-induced mutagenesis and carcinogenesis in skin. *Photochem. Photobiol.*, 2007; 83: 196-204.

[35] Lao, V. V., J. L. Herring, C. H. Kim, A. Darwanto, U. Soto, L. C. Sowers, Incorporation of 5-chlorocytosine into mammalian DNA results in heritable gene silencing and altered cytosine methylation patterns. *Carcinogenesis*, 2009; 30: 886-893.

[36] Lee, D. H., G. P. Pfeifer, Deamination of 5-methylcytosines within cyclobutane pyrimidine dimers is an important component of UVB mutagenesis. *J. Biol. Chem.*, 2003; 278: 10314-10321.

[37] Moving AHEAD with an international human epigenome project. *Nature*, 2008; 454: 711-715.

[38] Arasaradnam, R. P., D. M. Commane, D. Bradburn, J. C. Mathers, A review of dietary factors and its influence on DNA methylation in colorectal carcinogenesis. *Epigenetics*, 2008; 3: 193-198.

[39] Bingham, S., E. Riboli, Diet and cancer--the European Prospective Investigation into Cancer and Nutrition. *Nat. Rev. Cancer*, 2004; 4: 206-215.

[40] Parasramka, M. A., E. Ho, D. E. Williams, R. H. Dashwood, MicroRNAs, diet, and cancer: New mechanistic insights on the epigenetic actions of phytochemicals. *Molecular Carcinogenesis*, 2011: n/a-n/a.

[41] Hurt, E. M., W. L. Farrar, Cancer stem cells: the seeds of metastasis? *Mol. Interv.*, 2008; 8: 140-142.

[42] Dashwood, R. H., E. Ho, Dietary histone deacetylase inhibitors: from cells to mice to man. *Semin. Cancer Biol.*, 2007; 17: 363-369.

[43] Dashwood, R. H., M. C. Myzak, E. Ho, Dietary HDAC inhibitors: time to rethink weak ligands in cancer chemoprevention? *Carcinogenesis*, 2006; 27: 344-349.

[44] Kawasaki, B. T., E. M. Hurt, T. Mistree, W.L. Farrar, Targeting cancer stem cells with phytochemicals. *Molecular interventions*, 2008; 8: 174-184.

[45] Vel Szic, K. S., M. N. Ndlovu, G. Haegeman, W. Vanden Berghe, Nature or nurture: let food be your epigenetic medicine in chronic inflammatory disorders. *Biochemical Pharmacology*, 2010; 80: 1816-1832.

[46] Szyf, M., The dynamic epigenome and its implications in toxicology. *Toxicol. Sci.*, 2007; 100: 7-23.

[47] Watson, R. E., J. I. Goodman, Epigenetics and DNA methylation come of age in toxicology. *Toxicol. Sci.*, 2002; 67: 11-16.

[48] Gronniger, E., B. Weber, O. Heil, N. Peters, F. Stab, H. Wenck, B. Korn, M. Winnefeld, F. Lyko, Aging and chronic sun exposure cause distinct epigenetic changes in human skin. *PLoS Genet.*, 2010; 6: e1000971.

[49] Meshorer, E., T. Misteli, Chromatin in pluripotent embryonic stem cells and differentiation. *Nat. Rev. Mol. Cell. Biol.*, 2006.

[50] Lowry, W. E., L. Richter, R. Yachechko, A. D. Pyle, J. Tchieu, R. Sridharan, A. T. Clark, K. Plath, Generation of human induced pluripotent stem cells from dermal fibroblasts. *Proc. Natl. Acad. Sci. US*, 2008; 105: 2883-2888.

[51] Zhu, X. Q., X. H. Pan, W. Wang, Q. Chen, R. Q. Pang, X. M. Cai, A. R. Hoffman, J. F. Hu, Transient in vitro epigenetic reprogramming of skin fibroblasts into multipotent cells. *Biomaterials*; 31: 2779-2787.

[52] Chung, H. M., C. H. Won, J. H. Sung, Responses of adipose-derived stem cells during hypoxia: enhanced skin-regenerative potential. *Expert Opin. Biol. Ther.*, 2009; 9: 1499-1508.

[53] Chi, P., C. D. Allis, G. G. Wang, Covalent histone modifications--miswritten, misinterpreted and mis-erased in human cancers. *Nat. Rev. Cancer*, 2010; 10: 457-469.

[54] Ernst, J., M. Kellis, Discovery and characterization of chromatin states for systematic annotation of the human genome. *Nature Biotechnology*, 2010; 28: 817-825.

[55] Ernst, J., P. Kheradpour, T. S. Mikkelsen, N. Shoresh, L. D. Ward, C. B. Epstein, X. Zhang, L. Wang, R. Issner, M. Coyne, M. Ku, T. Durham, M. Kellis, B. E. Bernstein, Mapping and analysis of chromatin state dynamics in nine human cell types. *Nature*, 2011; 473: 43-49.

[56] Perissi, V., K. Jepsen, C. K. Glass, M. G. Rosenfeld, Deconstructing repression: evolving models of co-repressor action. *Nat. Rev. Genet.*, 2010; 11: 109-123.

[57] Vire, E., C. Brenner, R. Deplus, L. Blanchon, M. Fraga, C. Didelot, L. Morey, A. Van Eynde, D. Bernard, J. M. Vanderwinden, M. Bollen, M. Esteller, L. Di Croce, Y. de Launoit, F. Fuks, The Polycomb group protein EZH2 directly controls DNA methylation. *Nature*, 2006; 439: 871-874.

[58] Brenner, C., R. Deplus, C. Didelot, A. Loriot, E. Vire, C. De Smet, A. Gutierrez, D. Danovi, D. Bernard, T. Boon, P. G. Pelicci, B. Amati, T. Kouzarides, Y. de Launoit, L. Di Croce, F. Fuks, Myc represses transcription through recruitment of DNA methyltransferase corepressor. *EMBO J.*, 2005; 24: 336-346.

[59] Hervouet, E., F. M. Vallette, P. F. Cartron, Dnmt3/transcription factor interactions as crucial players in targeted DNA methylation. *Epigenetics*, 2009; 4: 487-499.

[60] Gupta, R. A., N. Shah, K. C. Wang, J. Kim, H. M. Horlings, D. J. Wong, M. C. Tsai, T. Hung, P. Argani, J. L. Rinn, Y. Wang, P. Brzoska, B. Kong, R. Li, R. B. West, M. J. van de Vijver, S. Sukumar, H. Y. Chang, Long non-coding RNA HOTAIR reprograms chromatin state to promote cancer metastasis. *Nature*, 2010; 464: 1071-1076.

[61] Seet, B. T., I. Dikic, M. M. Zhou, T. Pawson, Reading protein modifications with interaction domains. *Nat. Rev. Mol. Cell. Biol.,* 2006; 7: 473-483.

[62] Fischle, W., Talk is cheap--cross-talk in establishment, maintenance, and readout of chromatin modifications. *Genes and Development*, 2008; 22: 3375-3382.

[63] Schreiber, S. L., B. E. Bernstein, Signaling network model of chromatin. *Cell,* 2002; 111: 771-778.

[64] Natoli, G., Maintaining Cell Identity through Global Control of Genomic Organization. *Immunity*, 2010; 33: 12-24.

[65] Natoli, G., S. Ghisletti, I. Barozzi, The genomic landscapes of inflammation. *Genes and Development*, 2011; 25: 101-106.

[66] Bickmore, W. A., B. Tolhuis, M. Blom, R. M. Kerkhoven, L. Pagie, H. Teunissen, M. Nieuwland, M. Simonis, W. de Laat, M. van Lohuizen, B. van Steensel, Interactions among Polycomb Domains Are Guided by Chromosome Architecture. *PLoS Genetics*, 2011; 7: e1001343.

[67] Van Steensel, B., Chromatin: constructing the big picture. *The EMBO Journal,* 2011; 30: 1885-1895.

[68] Esteller, M., Cancer epigenomics: DNA methylomes and histone-modification maps. *Nat. Rev. Genet.*, 2007; 8: 286-298.

[69] Bird, A., DNA methylation patterns and epigenetic memory. *Genes Dev.,* 2002; 16: 6-21.

[70] Jones, P. A., G. Liang, Rethinking how DNA methylation patterns are maintained. *Nat. Rev. Genet.*, 2009; 10: 805-811.

[71] Law, J. A., S. E. Jacobsen, Establishing, maintaining and modifying DNA methylation patterns in plants and animals. *Nat. Rev. Genet.*, 2010; 11: 204-220.

[72] Lees-Murdock, D. J., C. P. Walsh, DNA methylation reprogramming in the germ line. *Advances in Experimental Medicine and Biology*, 2008; 626: 1-15.

[73] Robertson, K. D., DNA methylation and human disease. Nature Reviews. *Genetics*, 2005; 6: 597-610.

[74] Miranda, T. B., P. A. Jones, DNA methylation: the nuts and bolts of repression. *Journal of Cellular Physiology*, 2007; 213: 384-390.

[75] Wang, L., N. C. Chia, X. Lu, D. M. Ruden, Hypothesis: Environmental regulation of 5-hydroxymethylcytosine by oxidative stress. *Epigenetics*, 2011; 6: 853-856.

[76] Perillo, B., M. N. Ombra, A. Bertoni, C. Cuozzo, S. Sacchetti, A. Sasso, L. Chiariotti, A. Malorni, C. Abbondanza, E. V. Avvedimento, DNA Oxidation as Triggered by H3K9me2 Demethylation Drives Estrogen-Induced Gene Expression. *Science*, 2008; 319: 202-206.

[77] Scarano, M. I., M. Strazzullo, M. R. Matarazzo, M. D'Esposito, DNA methylation 40 years later: Its role in human health and disease. *Journal of Cellular Physiology,* 2005; 204: 21-35.

[78] Lande-Diner, L., H. Cedar, Silence of the genes--mechanisms of long-term repression. Nature Reviews. *Genetics*, 2005; 6: 648-654.

[79] Vaissiere, T., C. Sawan, Z. Herceg, Epigenetic interplay between histone modifications and DNA methylation in gene silencing. *Mutation Research*, 2008; 659: 40-48.

[80] Lee, B. M., L. C. Mahadevan, Stability of histone modifications across mammalian genomes: implications for 'epigenetic' marking. *J. Cell. Biochem*., 2009; 108: 22-34.

[81] Chodavarapu, R. K., S. Feng, Y. V. Bernatavichute, P. Y. Chen, H. Stroud, Y. Yu, J. A. Hetzel, F. Kuo, J. Kim, S. J. Cokus, D. Casero, M. Bernal, P. Huijser, A. T. Clark, U. Kramer, S. S. Merchant, X. Zhang, S. E. Jacobsen, M. Pellegrini, Relationship between nucleosome positioning and DNA methylation. *Nature*, 2010; 466: 388-392.

[82] Cosgrove, M. S., C. Wolberger, How does the histone code work? *Biochemistry and Cell Biology*, 2005; 83: 468-476.

[83] Guil, S., M. Esteller, DNA methylomes, histone codes and miRNAs: tying it all together. *Int. J. Biochem. Cell. Biol*., 2009; 41: 87-95.

[84] Jenuwein, T., C. D. Allis, Translating the histone code. *Science*, 2001; 293: 1074-1080.

[85] Margueron, R., P. Trojer, D. Reinberg, The key to development: interpreting the histone code? *Curr. Opin. Genet. Dev*., 2005; 15: 163-176.

[86] Robertson, A. G., M. Bilenky, A. Tam, Y. Zhao, T. Zeng, N. Thiessen, T. Cezard, A. P. Fejes, E. D. Wederell, R. Cullum, G. Euskirchen, M. Krzywinski, I. Birol, M. Snyder, P. A. Hoodless, M. Hirst, M. A. Marra, S. J. Jones, Genome-wide relationship between histone H3 lysine 4 mono- and tri-methylation and transcription factor binding. *Genome Research*, 2008; 18: 1906-1917.

[87] Jones, R. S., Epigenetics: reversing the 'irreversible'. *Nature*, 2007; 450: 357-359.

[88] Perissi, V., M. G. Rosenfeld, Controlling nuclear receptors: the circular logic of cofactor cycles. *Nat. Rev. Mol. Cell. Biol*., 2005; 6: 542-554.

[89] Nunez, E., X. D. Fu, M. G. Rosenfeld, Nuclear organization in the 3D space of the nucleus - cause or consequence? *Curr. Opin. Genet. Dev*., 2009; 19: 424-436.

[90] Nolis, I. K., D. J. McKay, E. Mantouvalou, S. Lomvardas, M. Merika, D. Thanos, Transcription factors mediate long-range enhancer-promoter interactions. *Proceedings of the National Academy of Sciences of the United States of America,* 2009; 106: 20222-20227.

[91] Fujioka, M., X. Wu, J. B. Jaynes, A chromatin insulator mediates transgene homing and very long-range enhancer-promoter communication. *Development*, 2009; 136: 3077-3087.

[92] Metivier, R., G. Reid, F. Gannon, Transcription in four dimensions: nuclear receptor-directed initiation of gene expression. *EMBO Rep*., 2006; 7: 161-167.

[93] Lieberman-Aiden, E., N. L. van Berkum, L. Williams, M. Imakaev, T. Ragoczy, A. Telling, I. Amit, B. R. Lajoie, P. J. Sabo, M. O. Dorschner, R. Sandstrom, B. Bernstein, M. A. Bender, M. Groudine, A. Gnirke, J. Stamatoyannopoulos, L. A. Mirny, E. S. Lander, J. Dekker, Comprehensive mapping of long-range interactions reveals folding principles of the human genome. *Science*, 2009; 326: 289-293.

[94] Taft, R. J., K. C. Pang, T. R. Mercer, M. Dinger, J. S. Mattick, Non-coding RNAs: regulators of disease. *J. Pathol.*, 2009; 220: 126-139.

[95] Taft, R. J., E. A. Glazov, N. Cloonan, C. Simons, S. Stephen, G. J. Faulkner, T. Lassmann, A. R. Forrest, S. M. Grimmond, K. Schroder, K. Irvine, T. Arakawa, M. Nakamura, A. Kubosaki, K. Hayashida, C. Kawazu, M. Murata, H. Nishiyori, S. Fukuda, J. Kawai, C. O. Daub, D. A. Hume, H. Suzuki, V. Orlando, P. Carninci, Y. Hayashizaki, J. S. Mattick, Tiny RNAs associated with transcription start sites in animals. *Nat. Genet.*, 2009; 41: 572-578.

[96] Mattick, J. S., R. J. Taft, G. J. Faulkner, A global view of genomic information - moving beyond the gene and the master regulator. *Trends Genet.,* 2009.

[97] Mattick, J. S., P. P. Amaral, M. E. Dinger, T. R. Mercer, M. F. Mehler, RNA regulation of epigenetic processes. *Bioessays*, 2009; 31: 51-59.

[98] Tsai, M. C., O. Manor, Y. Wan, N. Mosammaparast, J. K. Wang, F. Lan, Y. Shi, E. Segal, H. Y. Chang, Long Noncoding RNA as Modular Scaffold of Histone Modification Complexes. *Science*, 2010.

[99] De Santa, F., I. Barozzi, F. Mietton, S. Ghisletti, S. Polletti, B. K. Tusi, H. Muller, J. Ragoussis, C. L. Wei, G. Natoli, A large fraction of extragenic RNA pol II transcription sites overlap enhancers. *PLoS Biology,* 2010; 8: e1000384.

[100] Lujambio, A., M. Esteller, How epigenetics can explain human metastasis: a new role for microRNAs. *Cell Cycle*, 2009; 8: 377-382.

[101] Lujambio, A., G. A. Calin, A. Villanueva, S. Ropero, M. Sanchez-Cespedes, D. Blanco, L. M. Montuenga, S. Rossi, M. S. Nicoloso, W. J. Faller, W. M. Gallagher, S. A. Eccles, C. M. Croce, M. Esteller, A microRNA DNA methylation signature for human cancer metastasis. *Proceedings of the National Academy of Sciences of the United States of America*, 2008; 105: 13556-13561.

[102] Lujambio, A., M. Esteller, CpG island hypermethylation of tumor suppressor microRNAs in human cancer. *Cell Cycle*, 2007; 6: 1455-1459.

[103] Denis, H., M. N. Ndlovu, F. Fuks, Regulation of mammalian DNA methyltransferases: a route to new mechanisms. *EMBO reports*, 2011; 12: 647-656.

[104] Ndlovu, M. N., H. Denis, F. Fuks, Exposing the DNA methylome iceberg. *Trends in Biochemical Sciences*, 2011.

[105] Mahfouz, M. M., RNA-directed DNA methylation: Mechanisms and functions. *Plant Signal Behav.*, 2010; 5.

[106] Medzhitov, R., T. Horng, Transcriptional control of the inflammatory response. *Nat. Rev. Immunol.*, 2009; 9: 692-703.

[107] Illi, B., C. Colussi, A. Grasselli, A. Farsetti, M. C. Capogrossi, C. Gaetano, NO sparks off chromatin: tales of a multifaceted epigenetic regulator. *Pharmacol. Ther.*, 2009; 123: 344-352.

[108] Paradise, W. A., B. J. Vesper, A. Goel, J. D. Waltonen, K. W. Altman, G. K. Haines, J. A. Radosevich, Nitric oxide: perspectives and emerging studies of a well known cytotoxin. *Int. J. Mol. Sci.*, 2010; 11: 2715-2745.

[109] Oeckinghaus, A., S. Ghosh, The NF-kappaB family of transcription factors and its regulation. *Cold Spring Harb. Perspect. Biol.*, 2009; 1: a000034.
[110] Hayden, M. S., S. Ghosh, Shared principles in NF-kappaB signaling. *Cell*, 2008; 132: 344-362.
[111] Allison, D. F., M. W. Mayo, StIKKing together: do multiple IKK pathways cooperate in the DNA-damage response? *Mol. Cell*, 2010; 37: 453-454.
[112] Renner, F., R. Moreno, M. L. Schmitz, SUMOylation-dependent localization of IKKepsilon in PML nuclear bodies is essential for protection against DNA-damage-triggered cell death. *Mol. Cell*, 2010; 37: 503-515.
[113] Viatour, P., M. P. Merville, V. Bours, A. Chariot, Phosphorylation of NF-kappaB and IkappaB proteins: implications in cancer and inflammation. *Trends Biochem. Sci.*, 2005; 30: 43-52.
[114] Clement, J. F., S. Meloche, M. J. Servant, The IKK-related kinases: from innate immunity to oncogenesis. *Cell Res.*, 2008; 18: 889-899.
[115] Greten, F. R., M. Karin, The IKK/NF-kappaB activation pathway-a target for prevention and treatment of cancer. *Cancer Lett.*, 2004; 206: 193-199.
[116] Vermeulen, L., W. V. Berghe, I. M. Beck, K. De Bosscher, G. Haegeman, The versatile role of MSKs in transcriptional regulation. *Trends Biochem. Sci.*, 2009; 34: 311-318.
[117] Chen, L. F., W. C. Greene, Shaping the nuclear action of NF-kappaB. *Nat. Rev. Mol. Cell. Biol.*, 2004; 5: 392-401.
[118] Chen, L. F., S. A. Williams, Y. Mu, H. Nakano, J. M. Duerr, L. Buckbinder, W. C. Greene, NF-kappaB RelA phosphorylation regulates RelA acetylation. *Mol. Cell. Biol.*, 2005; 25: 7966-7975.
[119] Edmunds, J. W., L. C. Mahadevan, Cell signaling. Protein kinases seek close encounters with active genes. *Science*, 2006; 313: 449-451.
[120] Zenz, R., R. Eferl, C. Scheinecker, K. Redlich, J. Smolen, H. B. Schonthaler, L. Kenner, E. Tschachler, E. F. Wagner, Activator protein 1 (Fos/Jun) functions in inflammatory bone and skin disease. *Arthritis Res. Ther.*, 2008; 10: 201.
[121] Zenz, R., E. F. Wagner, Jun signalling in the epidermis: From developmental defects to psoriasis and skin tumors. *Int. J. Biochem. Cell. Biol.*, 2006; 38: 1043-1049.
[122] Zoumpourlis, V., P. Papassava, S. Linardopoulos, D. Gillespie, A. Balmain, A. Pintzas, High levels of phosphorylated c-Jun, Fra-1, Fra-2 and ATF-2 proteins correlate with malignant phenotypes in the multistage mouse skin carcinogenesis model. *Oncogene*, 2000; 19: 4011-4021.
[123] Belguise, K., N. Kersual, F. Galtier, D. Chalbos, FRA-1 expression level regulates proliferation and invasiveness of breast cancer cells. *Oncogene*, 2005; 24: 1434-1444.
[124] Eferl, R., E. F. Wagner, AP-1: a double-edged sword in tumorigenesis. *Nat. Rev. Cancer,* 2003; 3: 859-868.
[125] Ozanne, B. W., H. J. Spence, L. C. McGarry, R. F. Hennigan, Invasion is a genetic program regulated by transcription factors. *Curr. Opin. Genet. Dev.*, 2006; 16: 65-70.
[126] Mantovani, A., P. Allavena, A. Sica, F. Balkwill, Cancer-related inflammation. *Nature*, 2008; 454: 436-444.
[127] Balkwill, F., A. Mantovani, Cancer and inflammation: implications for pharmacology and therapeutics. *Clin. Pharmacol. Ther.*, 2010; 87: 401-406.

[128] Balkwill, F., K. A. Charles, A. Mantovani, Smoldering and polarized inflammation in the initiation and promotion of malignant disease. *Cancer Cell*, 2005; 7: 211-217.

[129] DeNardo, D. G., L. M. Coussens, Inflammation and breast cancer. Balancing immune response: crosstalk between adaptive and innate immune cells during breast cancer progression. *Breast Cancer Res.*, 2007; 9: 212.

[130] Chang, H. Y., D. S. Nuyten, J. B. Sneddon, T. Hastie, R. Tibshirani, T. Sorlie, H. Dai, Y. D. He, L. J. van't Veer, H. Bartelink, M. van de Rijn, P. O. Brown, M. J. van de Vijver, Robustness, scalability, and integration of a wound-response gene expression signature in predicting breast cancer survival. *Proc. Natl. Acad. Sci. US*, 2005; 102: 3738-3743.

[131] Shaw, T. J., P. Martin, Wound repair at a glance. *J. Cell Sci.,* 2009; 122: 3209-3213.

[132] Schafer, M., S. Dutsch, U. auf dem Keller, F. Navid, A. Schwarz, D. A. Johnson, J. A. Johnson, S. Werner, Nrf2 establishes a glutathione-mediated gradient of UVB cytoprotection in the epidermis. *Genes Dev.*, 2010; 24: 1045-1058.

[133] Saw, C. L., M. T. Huang, Y. Liu, T. O. Khor, A. H. Conney, A. N. Kong, Impact of Nrf2 on UVB-induced skin inflammation/ photoprotection and photoprotective effect of sulforaphane. *Mol. Carcinog.*, 2011; 50: 479-486.

[134] Beyer, T. A., U. Auf dem Keller, S. Braun, M. Schafer, S. Werner, Roles and mechanisms of action of the Nrf2 transcription factor in skin morphogenesis, wound repair and skin cancer. *Cell Death Differ.*, 2007; 14: 1250-1254.

[135] Natoli, G., S. Saccani, D. Bosisio, I. Marazzi, Interactions of NF-kappaB with chromatin: the art of being at the right place at the right time. *Nat. Immunol.*, 2005; 6: 439-445.

[136] Hager, G. L., J. G. McNally, T. Misteli, Transcription dynamics. *Mol. Cell*, 2009; 35: 741-753.

[137] Sung, M. H., L. Salvatore, R. De Lorenzi, A. Indrawan, M. Pasparakis, G. L. Hager, M. E. Bianchi, A. Agresti, Sustained oscillations of NF-kappaB produce distinct genome scanning and gene expression profiles. *PLoS One*, 2009; 4: e7163.

[138] van Uden, P., N. S. Kenneth, R. Webster, H. A. Muller, S. Mudie, S. Rocha, Evolutionary conserved regulation of HIF-1beta by NF-kappaB. *PLoS Genet.*, 2011; 7: e1001285.

[139] Rohwer, N., C. Dame, A. Haugstetter, B. Wiedenmann, K. Detjen, C. A. Schmitt, T. Cramer, Hypoxia-inducible factor 1alpha determines gastric cancer chemosensitivity via modulation of p53 and NF-kappaB. *PLoS ONE,* 2010; 5: e12038.

[140] Reuter, S., S. C. Gupta, M. M. Chaturvedi, B. B. Aggarwal, Oxidative stress, inflammation, and cancer: how are they linked? *Free Radic. Biol. Med.*, 2010; 49: 1603-1616.

[141] Dey, A., V. Tergaonkar, D. P. Lane, Double-edged swords as cancer therapeutics: simultaneously targeting p53 and NF-kappaB pathways. *Nat. Rev. Drug Discov.*, 2008; 7: 1031-1040.

[142] Suttana, W., S. Mankhetkorn, W. Poompimon, A. Palagani, S. Zhokhov, S. Gerlo, G. Haegeman, W. Berghe, Differential chemosensitization of P-glycoprotein overexpressing K562/Adr cells by withaferin A and Siamois polyphenols. *Molecular Cancer*, 2010; 9: 99.

[143] Garcia-Rivera, D., R. Delgado, N. Bougarne, G. Haegeman, W. V. Berghe, Gallic acid indanone and mangiferin xanthone are strong determinants of immunosuppressive anti-

tumour effects of Mangifera indica L. bark in MDA-MB231 breast cancer cells. *Cancer Lett.*, 2011; 305: 21-31.

[144] Surh, Y. J., Cancer chemoprevention with dietary phytochemicals. *Nat. Rev. Cancer*, 2003; 3: 768-780.

[145] Li, J. W. H., J. C. Vederas, Drug Discovery and Natural Products: End of an Era or an Endless Frontier? *Science*, 2009; 325: 161-165.

[146] Harvey, A., Natural products in drug discovery. *Drug Discovery Today*, 2008; 13: 894-901.

[147] Rosenfeld, M. G., V. V. Lunyak, C. K. Glass, Sensors and signals: a coactivator/corepressor/epigenetic code for integrating signal-dependent programs of transcriptional response. *Genes Dev.*, 2006; 20: 1405-1428.

[148] Ghosh, S., M. S. Hayden, New regulators of NF-kappaB in inflammation. *Nat. Rev. Immunol.*, 2008; 8: 837-848.

[149] Dong, J., E. Jimi, H. Zhong, M. S. Hayden, S. Ghosh, Repression of gene expression by unphosphorylated NF-kappaB p65 through epigenetic mechanisms. *Genes and Development*, 2008; 22: 1159-1173.

[150] Vanden Berghe, W., M. N. Ndlovu, R. Hoya-Arias, N. Dijsselbloem, S. Gerlo, G. Haegeman, Keeping up NF-kappaB appearances: epigenetic control of immunity or inflammation-triggered epigenetics. *Biochem. Pharmacol.*, 2006; 72: 1114-1131.

[151] Bracken, A. P., K. Helin, Polycomb group proteins: navigators of lineage pathways led astray in cancer. *Nat. Rev. Cancer*, 2009; 9: 773-784.

[152] Gehani, S. S., S. Agrawal-Singh, N. Dietrich, N. S. Christophersen, K. Helin, K. Hansen, Polycomb group protein displacement and gene activation through MSK-dependent H3K27me3S28 phosphorylation. *Molecular Cell*, 2010; 39: 886-900.

[153] Ndlovu, N., C. Van Lint, K. Van Wesemael, P. Callebert, D. Chalbos, G. Haegeman, W. Vanden Berghe, Hyperactivated NF-{kappa}B and AP-1 transcription factors promote highly accessible chromatin and constitutive transcription across the interleukin-6 gene promoter in metastatic breast cancer cells. *Mol. Cell. Biol.*, 2009; 29: 5488-5504.

[154] Vanden Berghe, W., K. De Bosscher, E. Boone, S. Plaisance, G. Haegeman, The nuclear factor-kappaB engages CBP/p300 and histone acetyltransferase activity for transcriptional activation of the interleukin-6 gene promoter. *The Journal of biological chemistry*, 1999; 274: 32091-32098.

[155] Vermeulen, L., G. De Wilde, P. Van Damme, W. Vanden Berghe, G. Haegeman, Transcriptional activation of the NF-kappaB p65 subunit by mitogen- and stress-activated protein kinase-1 (MSK1). *EMBO J.*, 2003; 22: 1313-1324.

[156] Dawson, M. A., A. J. Bannister, B. Göttgens, S. D. Foster, T. Bartke, A. R. Green, T. Kouzarides, JAK2 phosphorylates histone H3Y41 and excludes HP1α from chromatin. *Nature*, 2009; 461: 819-822.

[157] Natoli, G., F. De Santa, Shaping alternative NF-kappaB-dependent gene expression programs: new clues to specificity. *Cell Death Differ.*, 2006; 13: 693-696.

[158] Ford, E., D. Thanos, The transcriptional code of human IFN-beta gene expression. *Biochim. Biophys. Acta*, 2010.

[159] Ramirez-Carrozzi, V. R., D. Braas, D. M. Bhatt, C. S. Cheng, C. Hong, K. R. Doty, J. C. Black, A. Hoffmann, M. Carey, S. T. Smale, A unifying model for the selective

regulation of inducible transcription by CpG islands and nucleosome remodeling. *Cell*, 2009; 138: 114-128.

[160] Ramirez-Carrozzi, V. R., A. A. Nazarian, C. C. Li, S. L. Gore, R. Sridharan, A. N. Imbalzano, S. T. Smale, Selective and antagonistic functions of SWI/SNF and Mi-2beta nucleosome remodeling complexes during an inflammatory response. *Genes Dev.*, 2006; 20: 282-296.

[161] Puto, L. A., J. C. Reed, Daxx represses RelB target promoters via DNA methyltransferase recruitment and DNA hypermethylation. *Genes Dev.*, 2008; 22: 998-1010.

[162] Maeda, G., T. Chiba, S. Kawashiri, T. Satoh, K. Imai, Epigenetic inactivation of IkappaB Kinase-alpha in oral carcinomas and tumor progression. *Clinical Cancer Research*, 2007; 13: 5041-5047.

[163] O'Gorman, A., A. Colleran, A. Ryan, J. Mann, L. J. Egan, Regulation of NF-{kappa}B responses by epigenetic suppression of I{kappa}B{alpha} expression in HCT116 intestinal epithelial cells. *American Journal of Physiology. Gastrointestinal and Liver Physiology*, 2010.

[164] Ashall, L., C. A. Horton, D. E. Nelson, P. Paszek, C. V. Harper, K. Sillitoe, S. Ryan, D. G. Spiller, J. F. Unitt, D. S. Broomhead, D. B. Kell, D. A. Rand, V. See, M. R. White, Pulsatile stimulation determines timing and specificity of NF-kappaB-dependent transcription. *Science*, 2009; 324: 242-246.

[165] El Gazzar, M., B. K. Yoza, X. Chen, B. A. Garcia, N. L. Young, C. E. McCall, Chromatin-specific remodeling by HMGB1 and linker histone H1 silences proinflammatory genes during endotoxin tolerance. *Mol. Cell. Biol.*, 2009; 29: 1959-1971.

[166] El Gazzar, M., B. K. Yoza, J. Y. Hu, S. L. Cousart, C. E. McCall, Epigenetic silencing of tumor necrosis factor alpha during endotoxin tolerance. *The Journal of biological chemistry*, 2007; 282: 26857-26864.

[167] El-Osta, A., D. Brasacchio, D. Yao, A. Pocai, P. L. Jones, R. G. Roeder, M. E. Cooper, M. Brownlee, Transient high glucose causes persistent epigenetic changes and altered gene expression during subsequent normoglycemia. *J. Exp. Med.*, 2008; 205: 2409-2417.

[168] Jirtle, R. L., M. K. Skinner, Environmental epigenomics and disease susceptibility. *Nat. Rev. Genet.*, 2007; 8: 253-262.

[169] Backdahl, L., A. Bushell, S. Beck, Inflammatory signalling as mediator of epigenetic modulation in tissue-specific chronic inflammation. *Int. J. Biochem. Cell. Biol.*, 2009; 41: 176-184.

[170] McGowan, P. O., M. J. Meaney, M. Szyf, Diet and the epigenetic (re)programming of phenotypic differences in behavior. *Brain Res.,* 2008; 1237: 12-24.

[171] Mariman, E. C., Epigenetic manifestations in diet-related disorders. *J. Nutrigenet. Nutrigenomics*, 2008; 1: 232-239.

[172] Dobosy, J. R., V. X. Fu, J. A. Desotelle, R. Srinivasan, M. L. Kenowski, N. Almassi, R. Weindruch, J. Svaren, D. F. Jarrard, A methyl-deficient diet modifies histone methylation and alters Igf2 and H19 repression in the prostate. *Prostate*, 2008; 68: 1187-1195.

[173] Junien, C., Impact of diets and nutrients/drugs on early epigenetic programming. *J. Inherit. Metab. Dis.*, 2006; 29: 359-365.

[174] Delage, B., R. H. Dashwood, Dietary manipulation of histone structure and function. *Annu. Rev. Nutr.*, 2008; 28: 347-366.

[175] Waterland, R. A., M. Travisano, K.G. Tahiliani, M.T. Rached, S. Mirza, Methyl donor supplementation prevents transgenerational amplification of obesity. *Int. J. Obes.* (Lond.), 2008; 32: 1373-1379.

[176] Dolinoy, D. C., J. R. Weidman, R. A. Waterland, R. L. Jirtle, Maternal genistein alters coat color and protects Avy mouse offspring from obesity by modifying the fetal epigenome. *Environ. Health Perspect.*, 2006; 114: 567-572.

[177] Waterland, R. A., R. L. Jirtle, Transposable elements: targets for early nutritional effects on epigenetic gene regulation. *Mol. Cell. Biol.*, 2003; 23: 5293-5300.

[178] Manach, C., J. L. Donovan, Pharmacokinetics and metabolism of dietary flavonoids in humans. *Free Radic. Res.*, 2004; 38: 771-785.

[179] Manach, C., A. Mazur, A. Scalbert, Polyphenols and prevention of cardiovascular diseases. *Curr. Opin. Lipidol.*, 2005; 16: 77-84.

[180] Manach, C., G. Williamson, C. Morand, A. Scalbert, C. Remesy, Bioavailability and bioefficacy of polyphenols in humans. I. Review of 97 bioavailability studies. *Am. J. Clin. Nutr.*, 2005; 81: 230S-242S.

[181] Manach, C., A. Scalbert, C. Morand, C. Remesy, L. Jimenez, Polyphenols: food sources and bioavailability. *Am. J. Clin. Nutr.*, 2004; 79: 727-747.

[182] Williamson, G., C. Manach, Bioavailability and bioefficacy of polyphenols in humans. II. Review of 93 intervention studies. *Am. J. Clin. Nutr.*, 2005; 81: 243S-255S.

[183] Nian, H., B. Delage, E. Ho, R. H. Dashwood, Modulation of histone deacetylase activity by dietary isothiocyanates and allyl sulfides: studies with sulforaphane and garlic organosulfur compounds. *Environ. Mol. Mutagen.*, 2009; 50: 213-221.

[184] Vaquero, A., D. Reinberg, Calorie restriction and the exercise of chromatin. *Genes Dev.*, 2009; 23: 1849-1869.

[185] Molinie, B., P. Georgel, Genetic and epigenetic regulations of prostate cancer by genistein. *Drug News Perspect.*, 2009; 22: 247-254.

[186] Allard, J. S., E. Perez, S. Zou, R. de Cabo, Dietary activators of Sirt1. *Mol. Cell. Endocrinol.*, 2009; 299: 58-63.

[187] Ladurner, A. G., Chromatin places metabolism center stage. *Cell*, 2009; 138: 18-20.

[188] Lee, W. J., J. Y. Shim, B. T. Zhu, Mechanisms for the inhibition of DNA methyltransferases by tea catechins and bioflavonoids. *Mol. Pharmacol.*, 2005; 68: 1018-1030.

[189] Zheng, Y. G., J. Wu, Z. Chen, M. Goodman, Chemical regulation of epigenetic modifications: opportunities for new cancer therapy. *Med. Res. Rev.*, 2008; 28: 645-687.

[190] Wigle, T. J., J. M. Herold, G. A. Senisterra, M. Vedadi, D. B. Kireev, C. H. Arrowsmith, S. V. Frye, W. P. Janzen, Screening for inhibitors of low-affinity epigenetic peptide-protein interactions: an AlphaScreen-based assay for antagonists of methyl-lysine binding proteins. *J. Biomol. Screen*; 15: 62-71.

[191] Luo, J., M. H. Kuo, Linking nutrient metabolism to epigenetics. *Cell Science Reviews,* 2009; 6: 49-54.

[192] Imai, S., L. Guarente, Ten years of NAD-dependent SIR2 family deacetylases: implications for metabolic diseases. *Trends Pharmacol. Sci.*, 2010; 31: 212-220.

[193] Chen, N. C., F. Yang, L. M. Capecci, Z. Gu, A. I. Schafer, W. Durante, X. F. Yang, H. Wang, Regulation of homocysteine metabolism and methylation in human and mouse tissues. *FASEB Journal*, 2010.

[194] Bistulfi, G., E. Vandette, S. Matsui, D. J. Smiraglia, Mild folate deficiency induces genetic and epigenetic instability and phenotype changes in prostate cancer cells. *BMC Biol.*, 2010; 8: 6.

[195] Ghoshal, K., X. Li, J. Datta, S. Bai, I. Pogribny, M. Pogribny, Y. Huang, D. Young, S. T. Jacob, A folate- and methyl-deficient diet alters the expression of DNA methyltransferases and methyl CpG binding proteins involved in epigenetic gene silencing in livers of F344 rats. *J. Nutr.*, 2006; 136: 1522-1527.

[196] Kim, J. M., K. Hong, J. H. Lee, S. Lee, N. Chang, Effect of folate deficiency on placental DNA methylation in hyperhomocysteinemic rats. *J. Nutr. Biochem.*, 2009; 20: 172-176.

[197] Ulrich, C. M., M. C. Reed, H. F. Nijhout, Modeling folate, one-carbon metabolism, and DNA methylation. *Nutr. Rev.*, 2008; 66 Suppl. 1: S27-30.

[198] Sharma, P., R. D. Senthilkumar, V. Brahmachari, E. Sundaramoorthy, A. Mahajan, A. Sharma, S. Sengupta, Mining literature for a comprehensive pathway analysis: a case study for retrieval of homocysteine related genes for genetic and epigenetic studies. *Lipids Health Dis.*, 2006; 5: 1.

[199] Pogribny, I. P., V. P. Tryndyak, L. Muskhelishvili, I. Rusyn, S. A. Ross, Methyl deficiency, alterations in global histone modifications, and carcinogenesis. *J. Nutr.*, 2007; 137: 216S-222S.

[200] Lee, D. H., D. R. Jacobs, Jr., M. Porta, Hypothesis: a unifying mechanism for nutrition and chemicals as lifelong modulators of DNA hypomethylation. *Environ. Health Perspect.*, 2009; 117: 1799-1802.

[201] Whittle, J. R., M. J. Powell, V. M. Popov, L. A. Shirley, C. Wang, R. G. Pestell, Sirtuins, nuclear hormone receptor acetylation and transcriptional regulation. *Trends Endocrinol. Metab.*, 2007; 18: 356-364.

[202] Imai, S., L. Guarente, Ten years of NAD-dependent SIR2 family deacetylases: implications for metabolic diseases. *Trends Pharmacol. Sci.*; 31: 212-220.

[203] Moreau, M., M. Neveu, S. Stephan, E. Noblesse, C. Nizard, N. S. Sadick, S. Schnebert, F. Bonte, M. Dumas, P. Andre, E. Perrier, Enhancing cell longevity for cosmetic application: a complementary approach. *J. Drugs Dermatol.*, 2007; 6: s14-19.

[204] Baur, J. A., Resveratrol, sirtuins, and the promise of a DR mimetic. *Mech. Ageing Dev.*; 131: 261-269.

[205] Mathers, J. C., Nutritional modulation of ageing: genomic and epigenetic approaches. *Mech. Ageing Dev.*, 2006; 127: 584-589.

[206] Mai, A., D. Cheng, M. T. Bedford, S. Valente, A. Nebbioso, A. Perrone, G. Brosch, G. Sbardella, F. De Bellis, M. Miceli, L. Altucci, epigenetic multiple ligands: mixed histone/protein methyltransferase, acetyltransferase, and class III deacetylase (sirtuin) inhibitors. *J. Med. Chem.*, 2008; 51: 2279-2290.

[207] Paul, A. T., V. M. Gohil, K. K. Bhutani, Modulating TNF-alpha signaling with natural products. *Drug Discov. Today*, 2006; 11: 725-732.

[208] Rios, J. L., M. C. Recio, J. M. Escandell, I. Andujar, Inhibition of transcription factors by plant-derived compounds and their implications in inflammation and cancer. *Curr. Pharm. Des.*, 2009; 15: 1212-1237.

[209] Deorukhkar, A., S. Krishnan, G. Sethi, B. B. Aggarwal, Back to basics: how natural products can provide the basis for new therapeutics. *Expert Opin. Investig. Drugs*, 2007; 16: 1753-1773.

[210] Khanna, D., G. Sethi, K. S. Ahn, M. K. Pandey, A. B. Kunnumakkara, B. Sung, A. Aggarwal, B. B. Aggarwal, Natural products as a gold mine for arthritis treatment. *Curr. Opin. Pharmacol.*, 2007; 7: 344-351.

[211] Bremner, P., M. Heinrich, Natural products as targeted modulators of the nuclear factor-kappaB pathway. *J. Pharm. Pharmacol.*, 2002; 54: 453-472.

[212] Karin, M., Y. Yamamoto, Q. M. Wang, The IKK NF-kappa B system: a treasure trove for drug development. *Nat. Rev. Drug. Discov.*, 2004; 3: 17-26.

[213] Fraga, M. F., M. Herranz, J. Espada, E. Ballestar, M. F. Paz, S. Ropero, E. Erkek, O. Bozdogan, H. Peinado, A. Niveleau, J. H. Mao, A. Balmain, A. Cano, M. Esteller, A mouse skin multistage carcinogenesis model reflects the aberrant DNA methylation patterns of human tumors. *Cancer Res.,* 2004; 64: 5527-5534.

[214] Li, Y., A. H. Sawalha, Q. Lu, Aberrant DNA methylation in skin diseases. *J. Dermatol. Sci.*, 2009; 54: 143-149.

[215] Millington, G. W., Epigenetics and dermatological disease. *Pharmacogenomics*, 2008; 9: 1835-1850.

[216] Howell, P. M., Jr., S. Liu, S. Ren, C. Behlen, O. Fodstad, A. I. Riker, Epigenetics in human melanoma. *Cancer Control*, 2009; 16: 200-218.

[217] Richards, H. W., E. E. Medrano, Epigenetic marks in melanoma. *Pigment. Cell Melanoma Res.*, 2009; 22: 14-29.

[218] Shaw, T., P. Martin, Epigenetic reprogramming during wound healing: loss of polycomb-mediated silencing may enable upregulation of repair genes. *EMBO Rep.,* 2009; 10: 881-886.

[219] Fuchs, E., V. Horsley, More than one way to skin. *Genes Dev.*, 2008; 22: 976-985.

[220] Shilo, S., S. Roy, S. Khanna, C. K. Sen, MicroRNA in cutaneous wound healing: a new paradigm. *DNA Cell. Biol.,* 2007; 26: 227-237.

[221] Ivan, M., The ongoing microRNA revolution and its impact in biology and medicine. *J. Cell. Mol. Med.*, 2008; 12: 1425.

[222] Yi, R., M. N. Poy, M. Stoffel, E. Fuchs, A skin microRNA promotes differentiation by repressing 'stemness'. *Nature*, 2008; 452: 225-229.

[223] Liu, J., K. M. Drescher, X. M. Chen, MicroRNAs and Epithelial Immunity. *Int. Rev. Immunol.*, 2009; 28: 139-154.

[224] Schneider, M. R., miRNAs as novel players in skin development, homeostasis, and disease. *Br. J. Dermatol.*, 2011.

[225] Joyce, C. E., X. Zhou, J. Xia, C. Ryan, B. Thrash, A. Menter, W. Zhang, A.M. Bowcock, Deep sequencing of small RNAs from human skin reveals major alterations in the psoriasis miRNAome. *Hum. Mol. Genet.*, 2011.

[226] Balasubramanian, S., K. Lee, G. Adhikary, R. Gopalakrishnan, E. A. Rorke, R. L. Eckert, The Bmi-1 polycomb group gene in skin cancer: regulation of function by (-)-epigallocatechin-3-gallate. *Nutr. Rev.*, 2008; 66 Suppl. 1: S65-68.

[227] Olaharski, A. J., J. Rine, B. L. Marshall, J. Babiarz, L. Zhang, E. Verdin, M. T. Smith, The flavoring agent dihydrocoumarin reverses epigenetic silencing and inhibits sirtuin deacetylases. *PLoS Genet.*, 2005; 1: e77.

[228] Meeran, S. M., T. Singh, T. R. Nagy, S. K. Katiyar, High-fat diet exacerbates inflammation and cell survival signals in the skin of ultraviolet B-irradiated C57BL/6 mice. *Toxicol. Appl. Pharmacol.*, 2009; 241: 303-310.

[229] Altucci, L., S. Minucci, Epigenetic therapies in haematological malignancies: searching for true targets. *Eur. J. Cancer*, 2009; 45: 1137-1145.

[230] Beck, I. M., W. V. Berghe, S. Gerlo, N. Bougarne, L. Vermeulen, K. De Bosscher, G. Haegeman, Glucocorticoids and mitogen- and stress-activated protein kinase 1 inhibitors: possible partners in the combat against inflammation. *Biochem. Pharmacol.*, 2009; 77: 1194-1205.

[231] Beck, I. M., W. Vanden Berghe, L. Vermeulen, N. Bougarne, B. Vander Cruyssen, G. Haegeman, K. De Bosscher, Altered subcellular distribution of MSK1 induced by glucocorticoids contributes to NF-kappaB inhibition. *Embo. J.*, 2008; 27: 1682-1693.

[232] Beck, I. M., W. Vanden Berghe, L. Vermeulen, K. R. Yamamoto, G. Haegeman, K. De Bosscher, Crosstalk in inflammation: the interplay of glucocorticoid receptor-based mechanisms and kinases and phosphatases. *Endocr. Rev.*, 2009; 30: 830-882.

[233] Ogawa, S., J. Lozach, C. Benner, G. Pascual, R. K. Tangirala, S. Westin, A. Hoffmann, S. Subramaniam, M. David, M. G. Rosenfeld, C. K. Glass, Molecular determinants of crosstalk between nuclear receptors and toll-like receptors. *Cell*, 2005; 122: 707-721.

[234] Biddie, S. C., S. John, G. L. Hager, Genome-wide mechanisms of nuclear receptor action. *Trends Endocrinol. Metab.*, 2010; 21: 3-9.

[235] Wang, Y., W. Fiskus, D. G. Chong, K. M. Buckley, K. Natarajan, R. Rao, A. Joshi, R. Balusu, S. Koul, J. Chen, A. Savoie, C. Ustun, A. P. Jillella, P. Atadja, R. L. Levine, K. N. Bhalla, Cotreatment with panobinostat and JAK2 inhibitor TG101209 attenuates JAK2V617F levels and signaling and exerts synergistic cytotoxic effects against human myeloproliferative neoplastic cells. *Blood*, 2009; 114: 5024-5033.

[236] Fiskus, W., Y. Wang, A. Sreekumar, K. M. Buckley, H. Shi, A. Jillella, C. Ustun, R. Rao, P. Fernandez, J. Chen, R. Balusu, S. Koul, P. Atadja, V. E. Marquez, K. N. Bhalla, Combined epigenetic therapy with the histone methyltransferase EZH2 inhibitor 3-deazaneplanocin A and the histone deacetylase inhibitor panobinostat against human AML cells. *Blood*, 2009; 114: 2733-2743.

[237] Fiskus, W., K. Buckley, R. Rao, A. Mandawat, Y. Yang, R. Joshi, Y. Wang, R. Balusu, J. Chen, S. Koul, A. Joshi, S. Upadhyay, P. Atadja, K. N. Bhalla, Panobinostat treatment depletes EZH2 and DNMT1 levels and enhances decitabine mediated de-repression of JunB and loss of survival of human acute leukemia cells. *Cancer Biol. Ther.*, 2009; 8: 939-950.

[238] Davalos, V., M. Esteller, MicroRNAs and cancer epigenetics: a macrorevolution. *Curr. Opin. Oncol.*; 22: 35-45.

In: Nutrigenomics
Editor: M. T. Ravi Subbiah
ISBN: 978-1-62417-051-5

Chapter IX

Use of Nanotechnology in the Development of Cosmeceuticals for Skin Health

Prapaporn Boonme,[1] Paola A. Rojas,[2] Tatiana Andreani,[3,4] Aristoteles Rosmaninho,[5] Johann W. Wiechers[6] and Eliana B. Souto[4,7*#]

[1]Department of Pharmaceutical Technology, Faculty of Pharmaceutical Sciences, Prince of Songkla University, Thailand
[2]Department of Chemistry, Faculty of Science, University of los Andes, Bogotá, Colombia
[3]Department of Biology and Environment, University of Tras-os Montes e Alto Douro, Vila Real, Portugal
[4]Faculty of Health Sciences, Fernando Pessoa University, Porto, Portugal
[5]Centro Hospitalar do Porto, EPE-HSA, Department of Dermatology, Porto, Portugal
[6]JW Solutions, Gouda, the Netherlands
[7]Institute of Biotechnology and Bioengineering, Centre of Genetics and Biotechnology - University of Trás-os-Montes and Alto Douro (CGB-UTAD/IBB),Vila Real, Portugal

* Corresponding authors: Prapaporn Boonme and Eliana B. Souto: Department of Pharmaceutical Technology, Faculty of Pharmaceutical Sciences, Prince of Songkla University, 15 Karnjanavanich Road, Hat-Yai, Songkhla 90112, Thailand; Faculty of Health Sciences, Fernando Pessoa University, Rua Carlos da Maia, Nr. 296, P-4200-150 Porto, Portugal.

Paola A. Rojas: Department of Chemistry, Faculty of Science, University of los Andes, Cr 1 No 18A-70, Bogotá, Colombia. Tatiana Andreani: Department of Biology and Environment, University of Tras-os Montes e Alto Douro, P.O. Box 1013, 5001-801, Vila Real, Portugal. Faculty of Health Sciences, Fernando Pessoa University, Rua Carlos da Maia, Nr. 296, P-4200-150 Porto, Portugal. Aristoteles Rosmaninho: Centro Hospitalar do Porto, EPE-HSA, Department of Dermatology, Rua D. Manuel II s/n, Edifício das Consultas Externas, 4099-001 Porto, Portugal. Johann W. Wiechers: JW Solutions, Gasthuispolderweg 30, 2807 LL Gouda, the Netherlands. Eliana B. Souto: Faculty of Health Sciences, Fernando Pessoa University, Rua Carlos da Maia, Nr. 296, P-4200-150 Porto, Portugal. Institute of Biotechnology and Bioengineering, Centre of Genetics and Biotechnology - University of Trás-os-Montes and Alto Douro (CGB-UTAD/IBB), P.O. Box 1013, 5001-801, Vila Real, Portugal.

Abstract

Healthy skin contributes not only to the appearance and body homeostasis of people, but also offers barrier protection against pathogen invasion and water loss. Since beauty plays a vital role in most people's lives, novel cosmetic products are always in the pipelines to meet the demand of the consumer. More effective products such as cosmeceuticals and skin nutraceuticals are gaining popularity among consumers. Many nutraceuticals and cosmeceuticals are known to interact with the genome and may therefore cause changes in the skin through altered gene expression. Finding new chemicals and suitable formulations is becoming even more challenging today. Widely investigated formulations include nanotechnology-based carriers such as nanoemulsions, microemulsions and lipid nanoparticles (solid lipid nanoparticles (SLN), nanostructured lipid carriers (NLC)). In this chapter, we describe the nature of cosmeceuticals and skin nutraceuticals, the rationale behind their choices, their formulation into nanotechnology based carriers and successful examples of their application for the delivery of active ingredients.

Keywords: cosmeceuticals, skin nutraceuticals, nanotechnology products, lipid nanocarriers, nanoemulsions, microemulsions, solid lipid nanoparticles (SLN), nanostrctured lipid carriers (NLC)

1. Introduction

Nanotechnology is a well-known term applied to the creation and/or manipulation of materials within the nanometer range and covering the control, design, production, characterization and applications of these nanosized materials [1]. Nanosized materials exhibit physicochemical properties which are usually very different from their bulk counterparts. Some of their features include changes in color, physical strength, conductivity (heat and electrical), chemical reactivity, magnetism and optical properties. Optimization of these physicochemical characteristics may enhance or improve properties in several of the nanosized materials [2].

Among the materials used for the delivery of active ingredients in cosmetic sciences, a large number of nanocarriers has been developed over the last couple of years, such as micelles, liposomes, nanoemulsions, microemulsions, and biopolymeric as well as lipid nanoparticles. These carriers have been exploited for delivery of several types of cosmeceuticals and skin nutraceuticals [3, 4].

Over the last decades, nanocarriers have also been tested for oral delivery of nutraceutics (e.g., retinol [5], omega fatty acids [6], lycopene, carotenes and other anti-oxidants, and probiotics), for food safety and biosecurity (e.g., nanosensors), for cell-wall rupture applications, for loading with deodorants, disinfectants, antimicrobials and antifungals [7], and also for drinking water purification [8].

The current chapter is a comprehensive review of the latest achievements of nanotechnology in the use of cosmeceuticals and nutraceuticals for skin health.

2. Definitions of Cosmeceuticals and Skin Nutraceuticals

2.1. Cosmeceuticals

Cosmeceuticals are an undefined category of dermatology and in many countries including US, this group of cosmetics does not have clearly defined legal guidelines. To better understand its concept, a better approach would be to define what a 'cosmetic product' and 'drugs' are, since cosmeceuticals are in a region bordering these two categories.

Cosmetic products are been defined by the European Directive 93/35 EEC [9] as "any substance or preparation intended to be placed in contact with various external parts of the human body (epidermis, hair system, nails, lips and external genital organs) or with the teeth and mucous membranes of the oral cavity with a view exclusively or mainly to cleaning them, perfuming them, changing their appearance and/or correcting body odors and/or protecting them or keeping them in good condition". Drugs, on the other hand, have been defined by the Food Drug and Cosmetic (FDandC) Act as "(a) articles intended for use in diagnosis, cure, mitigation, treatment, or prevention of disease and (b) articles (other than food) intended to affect the structure or any function of the body in man or other animals" [9].

Cosmeceuticals, a new term widely popularized by Albert Kligman, lie at the intersection between drugs and cosmetic products [10]. These products are reported to possess two different application potentials, i.e., to enhance health condition and improve beauty of the skin by external application of an active ingredient which affects the skin barrier and health. Also, it is expected that new cosmeceuticals will combine the aesthetics of a cosmetic with the efficacy of a 'cosmetically active' ingredient. Moreover, it must produce an immediate benefit like a cosmetic, and in parallel to provide a long-term cosmetic benefit like a drug [11]. Cosmeceuticals can therefore be described as multifunctional products that can be bought without a prescription (such as the well known over-the-counter (OTC) drugs), providing more functionalities apart from their ability to provide scent, clean and skin coloration.

The US and Thai governments recognize cosmeceuticals as cosmetics in general. Therefore cosmetic ingredients included in the formulations should be obtained from raw materials with generally recognized as safe (GRAS) status. The main sources of new raw materials for the development of cosmeceuticals are often (but not limited to) botanical sources, especially those from plant extracts which are considered to be safe by the Food and Drug Administration (FDA).

With respect to active ingredients used in cosmeceutical formulations, interesting examples are the alpha hydroxyacids (AHAs) which are also named as fruit acids since they exist in several fruits. The first cosmeceuticals used to improve cutaneous exfoliation were glycolic and lactic acids which represent the smallest AHAs. However, their low pH and the fast penetration into the skin have been associated to skin irritation. A more suitable cosmeceutical used for this purpose would be niacinamide (the amide form of vitamin B3) which improves the skin texture without degrading the barrier function [11]. Nevertheless, glycolic and lactic acids are still widely used in skin care products [12] with formulation designs that limit their irritation potential. These AHAs can be combined with polymers that block the shunt routes into the skin slowing down the penetration rate and thus reduce skin irritation. When AHAs like mandelic acid and benzilic acid are bound to a phenyl group their lipophilicity increases, which can be used to target oily and acne-prone skin. Citric acid, an AHA which can be classified as a beta hydroxyacid (BHA), is widely used in topical formulations as an antioxidant and a pH adjustor [12]. Gluconolactone is the most commercialized polyhydroxy acid (PHA) for skin care products due to its anti-aging and UV protection properties thereby preventing the increase in sunburn cells after UVB irradiation. It can also be used in formulations with oxidative compounds since it reduces irritation and erythema caused by oxidative drugs, such as benzoyl peroxide [13-16]. Peptides are also used as cosmeceuticals, such as palmitoyl-lysine-threonine-threonine-lysine-serine (pal-KTTKS) which can improve the barrier function and also diminish skin wrinkling. KTTKS is a fragment of human dermal collagen I, which has been tested for its ability to accelerate wound healing by stimulating new collagen production.

Also, there is a study reporting that synthetic pal-KTTKS reduces fine lines and wrinkles in photo-aged facial appearance [17]. Topical application of retinoids has been shown to improve skin wrinkling. These compounds reduce the visible wrinkles by enhancing epidermal proliferation, preventing oxidative stress, controlling cutaneous bacterial flora, and may also act as UV filters when loaded in lipid nanocarriers. However, several compounds from the retinoid family such as retinaldehyde and retinol (vitamin A) can cause skin irritation when they are used in concentrations higher than 1% [18].

Recent studies have demonstrated that retinyl retinoate is more effective on hyaluronan production and less irritant than other retinoids [19]. The amino-monosaccharide, N-acetyl glucosamine, is used to improve facial depigmentation. It is used as a substrate for the production of hyaluronic acid, heparin sulfate and proteoglycans. These compounds help maintain the water content of the dermis, and inhibit the glycolsylation of tyrosinase, an important enzyme in melanin production [20]. Facial depigmentation can also be enhanced with ingredients that prevent both UVA and UVB radiation. Organic sun filters, such as octocrylene, avobenzone, oxybenzone and ecamsule, are widely used in cosmetic formulations, providing UVA and UVB protection [21].

New ingredients for cosmeceutical products are continuously being identified. However, their regulatory approval system is not as strict as it is for the drugs. For instance, FDA has a voluntary Cosmetic Registration program in US On the other hand, for drugs it is mandatory to register new drugs via the regulatory systems. Therefore, if there is no complaint, cosmetic industries are not required to register their data of cosmetic ingredients or to report cosmetic ingredient-related injuries to FDA [22].

However, the cosmetic industry has adopted a self-policing safety approach towards their new products to avoid losing consumer trust and reputation of the industry. These voluntary studies are not always thorough enough. Clinical tests to prove the benefit of an active ingredient or its risks are not routinely carried out since this is only required when a product efficacy is claimed. For cosmetic testing, invasive measurements are often not necessary because these claims are limited only to cosmetic properties [11].

Some of the non-invasive techniques used to evaluate the efficacy of cosmetic products are: 1) evaporimetry, which measures trans epidermal water loss (TEWL); 2) corneometry, to determine the state of the skin moisture; 3) chromametry to study skin color; 4) cutometry, which is used to determine biomechanical properties of the skin such as the elasticity, firmness and fatigue resistance; and 5) profilometry, which is used to measure skin roughness; and scan ultrasound imaging which is used to determine skin thickness [11, 23].

Despite all the developments in the cosmetic industry, there is still a need to develop more guidelines for a 'clear definition' and classification for cosmetic ingredients that falls between drugs and cosmetics. This will help cosmetic companies to develop more efficacious products with more robust active ingredients that can satisfy the needs of the consumers.

2.2. Skin Nutraceuticals

The term "nutraceutical" is a combination of two words, namely, "nutrition" and "pharmaceutical" and was introduced by Stephen DeFelice in 1989 [24]. Nutraceuticals are essential components for a wide range of industrial products, including pharmaceutics, health-care products, cosmetics, agrochemicals, and foodstuff. Examples of bioactives used as nutraceuticals include drugs, vitamins, antimicrobials, antioxidants, probiotics, prebiotics, peptides and proteins, carotenoids, omega fatty acids, flavorings, colorants, and preservatives. These active ingredients can occur in a variety of different molecular and physical forms. Even though the involvement of these molecules in physiological functions is not yet fully understood, it is widely accepted that their addition to food systems may have physiological benefits and disease-preventing properties [25].

Several nutraceuticals have made claims for items related to beauty and skin appearance, e.g., body weight management, fat burning efficacy, cellulite reduction etc. The skin appearance may also be related to individual gene expression. In fact, the amount and type of melanin (the cause of the skin color) is controlled by four to six genes. Therefore, if nutraceuticals can modify these genes, they can induce change in skin's appearance. Since skin appearance is one of important factors related to personality, nutraceuticals for these purposes have become very popular. Generally, nutraceuticals are taken via the oral route. However, many researchers have tried to formulate them in topical products and companies have found a large consumer market for such products.

In other words, skin nutraceuticals are used more like cosmeceuticals. For some nutraceuticals, encapsulation in topical nanocarriers might be difficult because of their sticky or greasy properties. Optimal skin penetration might also be difficult because of the large molecular size of nutraceuticals. In addition, topical products such as creams, gels and lotions contain high amount of water, which may decrease the stability of nutraceuticals. This is illustrated by the lower amount of commercially available topical products when compared to oral nutraceuticals. However, many promising active ingredients have been identified in the last few years.

Andre et al. [26] disclosed cosmetic products containing kola extracts in free or liposome form for cellulite reduction. Courtin [27] identified a plant nutriuretic peptide (PNP) that could be used as a slimming cosmetic composition. Gupta [28] developed several formulations of topical nutraceuticals, e.g., hydroxycitric acid in the form of facial masks and serums. Hydroxycitric acid is a nutraceutical that inhibits enzyme citrate lyase, and may be used for its fat burning efficacy.

There are many other commercial products of topical nutraceuticals (e.g., glutathione for skin-whitening) constantly being developed. Nevertheless, scientific evidence for efficacy was found only in a few selected cases. A major problem for the delivery of active ingredients via the topical route is their large molecular size. To overcome this limitation, nanocarriers have been proposed. Most promising examples include the nanoemulsions, microemulsions, and lipid nanoparticles.

3. Lipid Nanocarriers

Lipid nanocarriers may be defined as having one structural dimension with a diameter smaller than 100 nm [29] and being composed of lipid materials that are found in human tissues [30]. These systems can be produced either by the precise configuration of single atoms (bottom up approach), or by the reduction and refinement of material to the nanoscale dimension (top down approach) [31].

Since the main aim of these carriers to deliver the active ingredient into the site of action, they are required to protect the entrapped molecules against chemical and/or biological degradation during processing, handling, storage, and administration. To deliver and target a specific site within the organism, the nanocarrier has to show ability of controlling the release of the active ingredient.

It should also be compatible with the other components included in the formulation to achieve physicochemical advantages and qualitative outcomes. As such, nanotechnology is expected to affect the entire supply of novel materials making them more appealing to the consumers. One way of enhancing the properties of active ingredients may be is to load them into nanocarriers followed by processing into a final topical dosage form such as creams or lotions [32].

Examples of lipid nanocarriers for cosmetic applications include nanoemulsions, microemulsions, and lipid nanoparticles (i.e., solid lipid nanoparticles or SLN and nanostructured lipid carriers or NLC) as summarized in Table 1. Detailed description of these systems is given in the following sections.

Table 1. Basic steps for preparing lipid nanocarriers used in cosmetic applications

Type of lipid nanocarriers	Basic steps in preparations
Nanoemulsions	Prepared from emulsification of water and oil phases using an o/w surfactant as stabilizer, with physical and/or chemical energy input, to obtain a dispersion containing nanodroplets.
Microemulsions	Prepared from simple mixing of all components (i.e., water phase, oil phase, surfactant and cosurfactant), in a suitable ratio obtained from a microemulsion region of a pseudoternary phase diagram.
Solid Lipid Nanoparticles (SLN)	Prepared from pure solid lipids or a mixture of solid lipids, which recrystallize after a melting process, followed by cooling in which nanoparticles are produced with a solid core covered by a layer of surfactant molecules.
Nanostructured Lipid Carriers (NLC)	Prepared from a mixture of solid and liquid lipids which solidify after a melting process without crystallizing, remaining in the amorphous state.

3.1. Cosmetic Nanoemulsions and Microemulsions

Nanoemulsions are defined by optically isotropic low viscosity and kinetically stable dispersions. They consist of oil and water stabilized by an interfacial film of surfactant molecules [33]. In some cases, physicochemical characterization of these systems are not easy due the complexity and variety of structures and components involved in the development of the nanoemulsion and the limitations of the analytical techniques available. Some of the methods to identify and characterize these nanocarriers include the viscosity, conductivity, nuclear magnetic resonance (NMR), differential scanning calorimetry (DSC), cryo-field emission scanning electron microscopy (cryo-FESEM) and fluorescence resonance energy transfer (FRET) like those for microemulsion characterization [34, 35].

One of the main advantages of nanoemulsions over conventional emulsions or microemulsions is the lower surfactant concentration required for their production. Nanoemulsions can be formed by emulsification with physical or chemical energy input. The last process is based on emulsification of the oil phase into nanodroplets, due to some kind of interfacial instability originating from the rapid diffusion of solvent molecules across the interface and a decrease of the interfacial tension. Moreover, nanoemulsions are extremely stable against sedimentation or creaming due to their small size causing the diffusion rate to be faster than sedimentation rate [36].

Several other features of nanoemulsions make them more attractive for application in cosmetic products. Due to their small size nanoemulsions can penetrate the "rough" skin surface with uniform deposition. In addition, the presence of a low oil concentration in the system produces a desirable aesthetic character and skin feel. They can also be used for fragrance delivery and can substitute liposomes in many personal care products [37, 38]. Although nanoemulsions provide great advantages as a delivery system, some limitations have to be considered such as expensive manufacturing process involved and instability over longer time periods [39].

Microemulsions have also been used for similar purposes. A typical example is the o/w microemulsion composed of soy lecithin and decylpolyglucose, which is used to obtain transparent vehicles for sunscreens with lower percentage of surfactants [40]. However, because of the low viscosity of microemulsions, these nanocarriers might not be suitable for direct skin application in some instances [41]. This limitation can be overcome by the addition of an appropriate polymer or thickening agent to modify its rheology without influencing other features of the microemulsion like stability and water to oil ratios [42]. Another advantage of microemulsions is that its structure allows high incorporation of oil-soluble, water-soluble or amphiphilic molecules [43]. As a result loading and dispersability properties are improved in comparison to the individual microemulsion components [44, 45]. It was also found that the higher solubilization of active ingredients in microemulsions can result in higher concentration gradient between the formulation and the skin and provide more thermodynamic activity and higher transdermal flux (when compared to conventional vehicles) [46]. In addition, the transport of caffeine to the hypodermis by an alcohol-free o/w microemulsion was 1.3-fold larger than that from the emulsion and gel dosage forms [47]. This can be explained by the solubilization favoring drug partition into skin by modification of the thermodynamic activity of the active ingredient into the stratum corneum [48]. The nano-sized droplets (usually within the range between 10 and 140 nm) dispersed in the continuous phase of microemulsions can move easily through the stratum corneum and carry the active ingredient through the skin barrier enhancing the penetration of the active ingredient [49, 50]. Moreover, surfactants used in the formulation can reduce the diffusional barrier of the stratum corneum improving the penetration [48, 51]. However, very high concentrations can have the opposite effect [52]. Additionally, microemulsions can act as active ingredient reservoirs where loaded active ingredient is released from the dispersed phase to the continuous phase and further to the skin [50]. The release of the active ingredient into the external phase is relatively fast because of the small particle size that yields a large interfacial area contacting the skin where the active ingredient can easily be absorbed [53]. However, that will also depend on the lipophilicity of the active ingredient. In general, nanoemulsions and microemulsions can protect the entrapped active ingredients against degradation from hydrolysis or oxidation [54, 55]. In some cases, one type of emulsion can stabilize an active ingredient to a higher extent than other types of emulsions. Ascorbyl palmitate was found to be more stable in w/o microemulsions than in o/w microemulsions because its cyclic ring (the group sensitive to oxidative degradation located in the water phase) was protected by the microemulsion droplets. Besides that, its stability also depends on the amount of oxygen dissolved in the system, the storage conditions (i.e., light incidence) and its final concentration in the formulation. On the other hand, an active ingredient like sodium ascorbyl phosphate was found to be stable in both types of microemulsions [56].

One of the difficulties to develop a suitable microemulsion formulation is the selection of the components since the activity of components itself depends on the structure of the microemulsion, the character of the active ingredient and the membrane properties. The same substances can on occasion act as an enhancer or as retarder [57]. For instance, substances which are known as skin penetration enhancers like ethanol, propylene glycol and phospholipids can also behave as retarders [58] because under certain circumstances they can reduce the partitioning into the skin. Thickening agents can also affect the active ingredient diffusion by acting as enhancers or retarders [59]. In w/o microemulsion of sodium ascorbyl phosphate thickened with xanthan gum, the amount of drug released decreased when compared to a non-thickened system because of the increased viscosity. In contrast, in the o/w microemulsion of sodium ascorbyl phosphate thickened with colloidal silica, the amount of drug released was higher than in a non-thickened system [42]. Another important feature to take into account is that the microemulsions need high concentration of surfactants and co-surfactants to reduce the interfacial tension between the oil and water phases as mentioned previously. Nevertheless, based on the hydrophilic lipophilic deviation - net average curvature (HLD-NAC) theory of Salager and Acosta, stable microemulsions can be obtained with less than 1% total surfactant and dispersing 50% oil in 50% water [60, 61]. This is important as high concentrations of these substances can cause skin irritation or toxicity although in general nonionic and zwitterionic surfactants are less toxic than ionic ones [62]. New raw materials are constantly being developed to reduce the toxicity, such as pentyl rhamnoside (a co-surfactant that is claimed to be used in the pseudophase diagram of water), isooctane and cetyl pyridinium bromide [63] and ethoxylated glycerides, which are all claimed to be able to act as biologically mild surfactants [64].

Although many application of microemulsions in cosmetics has been reported for skin care products [65, 66], skin-whitening products [67], personal products and anti-aging products [68], their use and popularity in the cosmetic market has been limited. This might be due to the need of high concentrations of surfactants and co-surfactants to obtain a stable microemulsion. Such high concentrations may cause an enhanced risk of skin irritation.

Systemic toxicity also has to be considered to exclude the possibility of the active absorption into the blood circulation. However, this should not be a major issue due to limited amounts (due to low penetration rates in comparison to oral route) that can reach systemic circulation. Further, wider use of criteria based on HLD-NAC theory such as the measurement of the characteristic curvatures (Cc's) and effective alkyl carbon numbers (EACNs) of emulsifiers and emollients will help avoid the need for higher level of surfactants.

3.2. Cosmetic Lipid Nanoparticles (SLN, NLC)

Lipid nanoparticles for cosmetic usage include the so-called solid lipid nanoparticles (SLN) and nanostructured lipid carriers (NLC) which are described as biodegradable, biotolerable and biocompatible nanocarriers [69]. Lipid nanoparticles consist of a lipid matrix, solid at body and room temperature but usually melting above 40ºC. SLN are prepared from solid lipids or a mixture of solid lipids which re-crystallize after a melting process. This is followed by a cooling peroid during which nanoparticles are produced with a solid core covered by a layer of surfactant molecules [69]. Alternatively, NLC are prepared from a mixture of solid and liquid lipids which solidify after the melting process without crystallizing and remain in the amorphous state. NLC are produced this way are also composed of a solid core, covered by an emulsifying agent. But the difference between the two is the lack of crystallization in the NLCs which allows higher drug loading capacity compared to SLNs [69,70].

One of the great advantages of SLN and NLC is that their composition is based on biodegradable and physiological materials like lipids and waxes (i.e., tristearin, cetyl palmitate). These excipients are commonly used in the pharmaceutical and cosmetic industries and are accepted by regulatory authorities for cutaneous applications [69]. Another advantage is that lipids found in these delivery systems can reduce the danger of allergic contact dermatitis that might be produced by the drug [71]. In addition, they can be prepared without using organic solvents and other potentially toxic additives there by reducing the toxicological risk [72].

Different production techniques can be used to produce lipid nanoparticles. Some of the most commonly used techniques include the high pressure homogenization (HPH) process under hot or cold conditions [73] and the microemulsion technique [74]. There are other less popular methods, i.e., the phase inversion-based technique [75], the use of a membrane contactor [76], the solvent emulsification-evaporation method [77] and the emulsification-diffusion technique [78].

Both SLN and NLC show several advantages for topical administration. The small size of the lipid particles ensures close contact with the stratum corneum and can increase the amount of the drug penetrating into the mucosa or skin [70]. The pores of the epidermis are of 3-10 μm size [79], and lipid particles were found to penetrate the skin remaining intact [80]. Also, the modulation of drug release is possible with SLN and NLC. Drug release from lipid particles occurs by diffusion and simultaneously the lipid particles are degraded in the body by enzymes [69]. Further advancements to control the drug release from these particles have been explored. For example, in the treatment of psoriasis an increase of drug transport has been observed after skin application. Cyclosporine-loaded NLC were incorporated into a cream with excess free cyclosporine creating a saturated system. When the drug was released from the particles a supersaturated system was created and skin penetration was improved. This release can be triggered by increasing the temperature or by water evaporation because it could induce a polymorphic transformation of the particle from a high disordered lipid structure to a more ordered stable modification, leading to an easier drug release via drug expulsion [81, 82].

These types of carriers can fit the drug within its internal structure depending on several factors that can influence the release of the drug. Therefore, depending upon the desired characteristics required for a given medical treatment, a SLN or NLC system can be designed. With respect to SLN, the design depends on the drug concentration in the melted lipids and on the production technique. There are three theoretical models for drug or active molecule distribution that can be obtained in the SLN, i.e., homogeneous matrix, active-free lipid core with active-enriched shell, and active-enriched core with active-free lipid shell [70, 81, 83]. For NLC, three incorporation models have also been described, depending on the type of lipid compound and the oil concentration used in their production [83]. NLC were reported to have better long-term stability than SLN because of highly concentrated dispersions. The dispersion presents a pearl-like network, and after dilution with fluids the network is destroyed releasing only single but not aggregated particles. However, in topical formulations the particle aggregation can influence entrapment of lipid nanoparticles in cream and hydrogels and in this way gelation can be reduced [84].

Moreover, another important feature of the lipid nanoparticles is their occlusive properties. Some fat(s) used in the particle production tend to adhere to cells and surfaces forming a monolayer film at the surface of the skin. This produces an occlusive effect that increases the skin hydration. This effect has been shown in Cutanvoa Nanorepair Q10 cream, a cosmetic product containing NLCs with the antioxidant coenzyme Q10 that is used for anti-aging purposes. This product was compared in an *in vivo* skin hydration study with an identical o/w cream without NLCs. It was found that the Cutanova Nanorepair Q10 cream increased the skin hydration to a higher extent than the cream without NLCs even after six weeks following application. This result can be due to the occlusive properties or due to the changes in the polarity of the stratum corneum especially facilitating the partitioning of the more hydrophilic penetrants [83].

SLN and NLC also show the ability to stabilize chemically labile drugs against hydrolysis and/or oxidation. The chemical stability of coenzyme Q10 was compared when loaded in NLC and in nanoemulsion dispersions. Coenzyme Q10 stability was higher when formulated in NLC than in nanoemulsions. After 28 day of light exposure, a higher percentage of coenzyme Q10 was maintained in NLC than in nanoemulsions. On the other hand, daylight exposure increases the viscosity of these systems due to gel formation. This phenomenon was also observed in SLN in which gel formation was found to be dependent on the light intensity [85]. To overcome this limitation, a good packaging design is required.

Lipid nanoparticles have also shown to have some limited sunscreen properties due to some scattering capability. When loaded with inorganic compounds, SLN and NLC have shown the ability to enhance the intrinsic solar protection factor (SPF) of sunscreens. This interesting effect was observed in some sunscreen formulations containing benzophenone-3 (BZ3) loaded in poly (ε-caprolactone) (PCL) and SLN. In this study, the SPF increases when BZ3 was encapsulated into both nanostructures. However, SLN of BZ3 exhibited lower toxicity in human keratinocytes and fibroblasts when compared to PCL formulation [86]. This characteristic of lipid nanoparticles is useful for the design of novel sunscreens. It helps to decrease the use of organic filters that may cause irritation and also may reduce the whitening effect caused by the use of large quantities of large inorganic sun filter particles.

4. Nutraceutical Nanocarriers for Skin Health in Marketed Formulations

A variety of nanotechnology-based products are currently available to provide protective barriers, flavour masking, controlled release, and better dispersibility for water-insoluble ingredients and additives. Nanocarriers containing active ingredients (e.g., probiotics, prebiotics, antioxidants, minerals, and other micronutrients) can be used to protect them from oxidation. This helps to improve their shelf life, getting to the taste receptor sites and thus to reduce the undesirable organoleptic characteristics in the finished product.

To improve bioavailability, nanotechnology renders hydrophilic substances fat soluble and liphophilic ones water soluble thereby allowing nanostructures of some nutraceuticals (e.g., carotenoids, phytosterols, and antioxidants) to be dispersed in soft formulations.

Nanocarriers of mean particle size of 100 nm loaded with lycopene have been reported by the FDA as a GRAS substance [87]. Another example of nanostructures for nutraceuticals is coenzyme Q10, which is an antioxidant agent with well-established pharmacological activities. However, coenzyme Q10 is only marketed as a nutritional supplement without any claim of its therapeutic activity probably due to its poor physicochemical and biopharmaceutical properties leading to low oral bioavailability. Ankola et al. reported advantages of nanostructures in improving the therapeutic value of molecules such as coenzyme Q10, facilitating its usage as first line agent for prophylaxis and therapy by overcoming the problems associated with its delivery in physiological conditions [88].

There are many other nutraceuticals that are poorly absorbed, such as vitamin B12, vitamin K2, vitamin E and many other phytonutrients such as the polyphenols and terpenes (e.g., carotenoids, chromonols, lemonoids, and saponins) in which nanotechnology has great potential to enhance their bioavailability and to generate new nutraceutical products. However, most of new nutraceutical products are used via oral intake rather than as topical products.

Conclusion

Cosmeceuticals are chemicals having an activity higher than ordinary ingredients used in cosmetics but lower than drugs. Nowadays, they are very popular since they respond to the higher demands of consumers for skin benefits such as skin-whitening, anti-wrinkle, or cellulite-reducing effects. Nutraceuticals are chemicals having a functionality half-way between nutrients and drugs. In general, nutraceuticals are administered via the oral route. Many nutraceuticals have also proven activity related to beauty and appearance such as enhancement of body fat burning and altering melanin production in white skin. Therefore, a topical route of administration seems only logical. The mechanisms of action of the new ingredients are varied depending on their specific properties. Of particular interest is the fact that some of them are capable of affecting gene expression in the skin, as well as in improving cutaneous exfoliation and scavenging free radicals. Due to the well-known health and beauty benefits attributed to cosmeceuticals and nutraceuticals, more sophisticated and efficient formulations for topical application are in great demand by the consumers. Nanotechnology formulations are poised to play an important role in cosmetic applications due to their ability to enhance product appearance, efficacy, stability, and keep up with the consumer demands.

References

[1] Mishra, B., Patel, B. B., Tiwari, S. Colloidal nanocarriers: a review on formulation technology, types and applications toward targeted drug delivery. *Nanomedicine*. 2010;6:9-24.

[2] Oberdörster, G. Safety assessment for nanotechnology and nanomedicine: concepts of nanotoxicology. *J. Intern. Med.* 2010;267:89-105.

[3] Chiang, H. M., Lin, J. W., Hsiao, P. L., Tsai, S. Y., Wen, K. C. Hydrolysates of citrus plants stimulate melanogenesis protecting against UV-induced dermal damage. *Phytother. Res.* 2011;25:569-576.

[4] Plonka, P. M. Electron paramagnetic resonance as a unique tool for skin and hair research. *Exp. Dermatol.* 2009;18:472-484.

[5] Shidhaye, S. S., Vaidya, R., Sutar, S., Patwardhan, A., Kadam, V. J. Solid lipid nanoparticles and nanostructured lipid carriers - innovative generations of solid lipid carriers. *Curr. Drug Deliv.* 2008;5:324-331.

[6] Souto, E. B., Müller, R. H., Gohla, S. A novel approach based on lipid nanoparticles (SLN) for topical delivery of alpha-lipoic acid. *J. Microencapsul.* 2005; 22:581-592.

[7] Weir, E., Lawlor, A., Whelan, A., Regan, F. The use of nanoparticles in anti-microbial materials and their characterization. *Analyst*. 2008;133:835-845.

[8] Harisha, R. S., Hosamani, K. M., Keri, R. S., Nataraj, S. K., Aminabhavi, T. M. Arsenic removal from drinking water using thin film composite nanofiltration membrane. *Desalination*. 2010;252:75-80.

[9] Council Directive 93/35/EEC amending for the sixth time Directive 76/768/EEC on the approximation of the laws of the Member States relating to cosmetic products. *Official J.* L151, p. 32-37.

[10] Kligman, A. *Cosmeceuticals*. In: Elsner, P., Maiback, H., editors. New York: Marcel Dekker, 2000. p. 1-8.

[11] Draelos, Z. D. Cosmeceuticals: undefined, unclassified, and unregulated. *Clin. Dermatol.* 2009;27:431-434.

[12] Green, B. A., Yu, R. J., Van Scott, E. J. Clinical and cosmeceutical uses of hydroxyacids. *Clin. Dermatol.* 2009;27:495-501.

[13] Berardesca, E., Distante, F., Vignoli, G. P., Oresajo, C., Green, B. Alpha hydroxyacids modulate stratum corneum barrier function. *Br. J. Dermatol.* 1997;137(6):934-938.

[14] Bernstein, E. F. Chemical peels. *Semin. Cutan. Med. Surg*. 2002;21:27-45.

[15] Bernstein, E. F., Lee, J., Brown, D. B., Yu, R., Van Scott, E. Glycolic acid treatment increases type I collagen mRNA and hyaluronic acid content of human skin. *Dermatol. Surg*. 2001;27:429-433.

[16] Bernstein, E. F., Andersen, D., Zelickson, B. D. Laser resurfacing for dermal photoaging. *Clin. Plast. Surg*. 2000;27:221-240.

[17] Lintner, K., Mas-Chamberlin, C., Mondon, P., Peschard, O., Lamy, L. Cosmeceuticals and active ingredients. *Clin. Dermatol.* 2009;27:461-468.

[18] Boehnlein, J., Sakr, A., Lichtin, J. L., Bronaugh, R. L. Characterization of esterase and alcohol dehydrogenase activity in skin: metabolism of retinyl palmitate to retinol (vitamin A) during percutaneous absorption. *Pharm. Res.* 1994;11:1155-1159.

[19] Kim, J. E., Kim, B., Kim, H., Kim, H., Lee, J. D., Kim, H. J., Choi, K. Y., Lee, S. H. Retinyl retinoate induces hyaluronan production and less irritation than other retinoids. *J. Dermatol.* 2010;37:448-454.

[20] Kimball, A. B., Kaczvinsky, J. R., Li, J., Robinson, L. R., Matts, P. J., Berge, C. A., Miyamoto, K., Bissett, D. L. Reduction in the appearance of facial hyperpigmentation after use of moisturizers with a combination of topical niacinamide and N-acetyl glucosamine: results of a randomized, double-blind, vehicle-controlled trial. *Br. J. Dermatol.* 2010;162:435-441.

[21] Beasley, D. G., Meyer, T. A. Characterization of the UVA protection provided by avobenzone, zinc oxide, and titanium dioxide in broad-spectrum sunscreen products. *Am. J. Clin. Dermatol.* 2010;11:413-421.

[22] Schueller, R., Romanowski, P. *Beginning Cosmetic Chemistry*. 3rd ed. IL: Allured Pub. Corp., 2009.

[23] Muggli, R. Systemic evening primrose oil improves the biophysical skin parameters of healthy adults. *Int. J. Cosmet. Sci.* 2005;27:243-249.

[24] Kalra, E. Nutraceutical - definition and introduction. *AAPS Pharm. Sci.* 2003;5:Article 25.

[25] Elliott, R., Ong, T. J. Science, medicine, and the future - nutritional genomics. *BMJ.* 2002;324(7351):1438-1442.

[26] Andre, P., Dominice, J., Perrier, P., Redziniak, G. Cosmetic or dermatological composition, containing kola extracts and method for treatment of cellulitis. US 005165935A, 1992.

[27] Courtin, O. Slimming cosmetic composition comprising a plant extract containing a plant nutriuretic peptide (PNP). US 2003/0007988 A1, 2003.

[28] Gupta, S. K. Topical nutraceutical compositions with selective body slimming and tone firming antiaging benefits. US 2004/0146539 A1, 2004.

[29] Prow, T. W., Grice, J. E., Lin, L. L., Faye, R., Butler, M., Becker, W., Wurm, E. M., Yoong, C., Robertson, T. A., Soyer, H. P., Roberts, M. S. Nanoparticles and microparticles for skin drug delivery. *Adv. Drug Deliv. Rev.* 2011;63:470-491.

[30] McNeil, S. E. Nanotechnology for the biologist. *J. Leukoc. Biol.* 2005;78:585-594.

[31] Silva, G. A. Neuroscience nanotechnology: progress, opportunities and challenges. *Nat. Rev. Neurosci.* 2006;7:65-74.

[32] Gulbake, A., Jain, A., Khare, P., Jain, S. K. Solid lipid nanoparticles bearing oxybenzone: in-vitro and in-vivo evaluation. *J. Microencapsul.* 2010;27:226-233.

[33] Shafiq, S., Shakeel, F., Talegaonkar, S., Ahmad, F. J., Khar, R. K., Ali, M. Development and bioavailability assessment of ramipril nanoemulsion formulation. *Eur. J. Pharm. Biopharm.* 2007;66:227-243.

[34] Boonme, P., Krauel, K., Graf, A., Rades, T., Junyaprasert, V. B. Characterization of microemulsion structures in the pseudoternary phase diagram of isopropyl palmitate/water/Brij 97:1-butanol. *AAPS Pharm. Sci. Tech.* 2006;7:Article 45.

[35] Boonme, P., Krauel, K., Graf, A., Rades, T., Junyaprasert, V. B. Characterisation of microstructures formed in isopropyl palmitate/water/Aerosol®OT:1-butanol (2:1) system. *Pharmazie*. 2006;61:927-932.

[36] Izquierdo, P., Esquena, J., Tadros, T. F., Dederen, J. C., Feng, J., Garcia-Celma, M. J., Azemar, N., Solans, C. Phase behavior and nano-emulsion formation by the phase inversion temperature method. *Langmuir*. 2004;20:6594-6598.

[37] Sonneville-Aubrun, O., Simonnet, J.-T., L'Alloret, F. Nanoemulsions: a new vehicle for skincare products. *Adv. Colloid Interface Sci.* 2004;108-109:145-149.

[38] Bouchemal, K., Briançon, S., Perrier, E., Fessi, H. Nano-emulsion formulation using spontaneous emulsification: solvent, oil and surfactant optimisation. *Int. J. Pharm.* 2004;280:241-251.

[39] Sharma, N., Bansal, M., Visht, S., Sharma, P. K., Kulkarni, G. T. Nanoemulsion: A new concept of delivery system. *Chron. Young Scientist.* 2010;1:2-6.

[40] Carlotti, M. E., Gallarate, M., Rossatto, V. O/W microemulsion as a vehicle for sunscreens. *J. Cosmet. Sci.* 2003;54:451-462.

[41] Gasco, M. R., Gallarate, M., Pattarino, F. In vitro permeation of azelaic acid from viscosized microemulsions. *Int. J. Pharm.* 1991;69:193-196.

[42] Spiclin, P., Homar, M., Zupancic-Valant, A., Gasperlin, M. Sodium ascorbyl phosphate in topical microemulsions. *Int. J. Pharm.* 2003; 256:65-73.

[43] He, C. X., He, Z. G., Gao, J. Q. Microemulsions as drug delivery systems to improve the solubility and the bioavailability of poorly water-soluble drugs. *Expert Opin. Drug Deliv.* 2010;7:445-460.

[44] Boonme, P. Applications of microemulsions in cosmetics. *J. Cosmet. Dermatol.* 2007;6:223-228.

[45] Junyaprasert, V. B., Boonme, P., Wurster, D. E., Rades, T. Aerosol OT microemulsions as carriers for transdermal delivery of hydrophobic and hydrophilic local anesthetics. *Drug Deliv.* 2008;15:323-330.

[46] Murthy, N. S., Shivakumar, H. N. Topical and Transdermal Drug Delivery. In: Kulkarni, V. S., Editor, *Handbook of Non-Invasive Drug Delivery Systems.* Boston: William Andrew Publishing, 2010. p. 1-36.

[47] Bolzinger, M. A., Briançon, S., Pelletier, J., Fessi, H., Chevalier, Y. Percutaneous release of caffeine from microemulsion, emulsion and gel dosage forms. *Eur. J. Pharm. Biopharm.* 2008;68:446-451.

[48] Delgado-Charro, M. B., Iglesias-Vilas, G., Blanco-Méndez, J., López-Quintela, M. A., Marty, J. P., Guy, R. H. Delivery of a hydrophilic solute through the skin from novel microemulsion systems. *Eur. J. Pharm. Biopharm.* 1997;43:37-42.

[49] Sintov, A. C., Botner, S. Transdermal drug delivery using microemulsion and aqueous systems: influence of skin storage conditions on the in vitro permeability of diclofenac from aqueous vehicle systems. *Int. J. Pharm.* 2006;311:55-62.

[50] Peltola, S., Saarinen-Savolainen, P., Kiesvaara, J., Suhonen, T. M., Urtti, A. Microemulsions for topical delivery of estradiol. *Int. J. Pharm.* 2003;254:99-107.

[51] Changez, M., Varshney, M. Aerosol-OT microemulsions as transdermal carriers of tetracaine hydrochloride. *Drug Dev. Ind. Pharm.* 2000;26:507-512.

[52] Ktistis, G., Niopas, I. A study on the in-vitro percutaneous absorption of propranolol from disperse systems. *J. Pharm. Pharmacol.* 1998;50:413-418.

[53] Lu, G. W., Gao, P. Emulsions and Microemulsions for Topical and Transdermal Drug Delivery. In: Kulkarni, V. S., Editor, *Handbook of Non-Invasive Drug Delivery Systems*. Boston: William Andrew Publishing, 2010. p. 59-94.

[54] Hagigit, T., Abdulrazik, M., Orucov, F., Valamanesh, F., Lambert, M., Lambert, G., Behar-Cohen, F., Benita, S. Topical and intravitreous administration of cationic nanoemulsions to deliver antisense oligonucleotides directed towards VEGF KDR receptors to the eye. *J. Control. Release.* 2010;145:297-305.

[55] Zhang, G., Liao, Y., Baker, Y. Surface engineering of core/shell iron/iron oxide nanoparticles from microemulsions for hyperthermia *Mater. Sci. Eng.* C. 2010;30:92-97.

[56] Spiclin, P., Gasperlin, M., Kmetec, V. Stability of ascorbyl palmitate in topical microemulsions. *Int. J. Pharm.* 2001;222:271-279.

[57] Kogan, A., Garti, N. Microemulsions as transdermal drug delivery vehicles. *Adv. Colloid. Interface Sci.* 2006;123-126:369-385.

[58] Kaushika, D., Costacheb, A., Michniak-Kohn, B. Percutaneous penetration modifiers and formulation effects. *Int. J. Pharm.* 2010;386:42-51.

[59] López, A., Llinares, F., Cortell, C., Herráez, M. Comparative enhancer effects of Span 20 with Tween 20 and Azone on the in vitro percutaneous penetration of compounds with different lipophilicities. *Int. J. Pharm.* 2000;202:133-140.

[60] Komesvarakul, N., Sanders, M. D., Szekeres, E., Acosta, E. J., Faller, J. F., Mentlik, T., Fisher, L. B., Nicoll, G., Sabatini, D. A., Scamehorn, J. F. Microemulsions of triglyceride-based oils: the effect of co-oil and salinity on phase diagrams. *J. Cosmet. Sci.* 2006;57:309-325.

[61] Szekeres, E., Acosta, E., Sabatini, D. A., Harwell, J. H. Modeling solubilization of oil mixtures in anionic microemulsions II: mixtures of polar and non-polar oils. *J. Colloid Interface Sci.* 2006;294:222-233.

[62] Kreilgaard, M. Influence of microemulsions on cutaneous drug delivery. *Adv. Drug Deliv. Rev.* 2002;54(Suppl. 1):S77-S98.

[63] Houlmont, J., Vercruysse, K., Perez, E., Rico-Lattes, I., Bordat, P., Treihou, M. Cosmetic use formulations containing pentyl rhamnoside and cetyl rhamnoside. *Int. J. Cosmet. Sci.* 2001;23:363-368.

[64] Gloukhova, T. V., Kienskaya, K. I., Myakon'kii, A. G., Kim, V. Formation and properties of microemulsions stabilized by ethoxylated glycerides. *Colloid J.* 2005;67(3):291-294.

[65] Guglielmini, G. Nanostructured novel carrier for topical application. *Clin. Dermatol.* 2008;26:341-346.

[66] Boonme, P. Uses of microemulsions as novel vehicles in skin care products. *HPC Today*. 2009;3:18-20.

[67] Boonme, P., Junyaprasert, V. B., Suksawad, N., Songkro, S. Microemulsions and nanoemulsions: novel vehicles for whitening cosmeceuticals. *J. Biomed. Nanotechnol.* 2009;5:373-383.

[68] Boonme, P., Yotsawimonwat, S. Anti-ageing microemulsions and nanoemulsions. *HPC Today*. 2011;5:42-46.

[69] Souto, E. B., Almeida, A. J., Müller, R. H. Lipid nanoparticles (SLN®, NLC®) for cutaneous drug delivery: structure, protection and skin effects. *J. Biomed. Nanotechnol.* 2007;3:317-331.

[70] Souto, E. B., Wissing, S. A., Barbosa, C. M., Muller, R. H. Development of a controlled release formulation based on SLN and NLC for topical clotrimazole delivery. *Int. J. Pharm.* 2004;278:71-77.

[71] Singh, K. K., Pople, P. Safer than safe: lipid nanoparticulate encapsulation of tacrolimus with enhanced targeting and improved safety for atopic dermatitis *J. Biomed. Nanotechnol.* 2011;7:40-41.

[72] Souto, E. B., Almeida, A. J., Müller, R. H. Feasibility studies of lipid-based carriers for dermal applications. *Proceedings of 14th International Workshop on Bioencapsulation and COST 865 Meeting*. Lausanne, Switzerland, 2006. p. 27.

[73] Severino, P., Pinho, S. C., Souto, E. B., Santana, M. H. A. Crystallinity of Dynasan®114 and Dynasan®118 matrices for the production of stable Miglyol®-loaded nanoparticles. *J. Therm. Anal. Calorim.* 2011;doi:10.1007/s10973-011-1613-7.

[74] Doktorovová, S., Araújo, J., Garcia, M. L., Rakovský, E., Souto, E. B. Formulating fluticasone propionate in novel PEG-containing nanostructured lipid carriers (PEG-NLC). *Colloids Surf. B Biointerfaces.* 2010;75:538-542.

[75] Heurtault, B., Saulnier, P., Pech, B., Proust, J. E., Benoit, J. P. A novel phase inversion-based process for the preparation of lipid nanocarriers. *Pharm. Res.* 2002;19:875-880.

[76] Li, Y., Fessi, H., Charcosset, C. Preparation of indomethacin-loaded lipid particles by membrane emulsification. *Adv. Sci. Lett.* 2011;4:591-595.

[77] Zhang, Q. X., Xiong, Q. P., Zhang, D. Y. Study on preparation and characterization of resveratrol solid lipid nanoparticles and its anticancer effects in vitro. *J. Chinese Medicinal Materials.* 2010;33:1929-1932. (English Abstract)

[78] Urbán-Morlán, Z., Ganem-Rondero, A., Melgoza-Contreras, L. M., Escobar-Chávez, J. J., Nava-Arzaluz, M. G., Quintanar-Guerrero, D. Preparation and characterization of solid lipid nanoparticles containing cyclosporine by the emulsification-diffusion method. *Int. J. Nanomedicine.* 2010;5:611-620.

[79] Rolland, A., Wagner, N., Chatelus, A., Shroot, B., Schaefer, H. Site-specific drug delivery to pilosebaceous structures using polymeric microspheres. *Pharm. Res.* 1993;10:1738-1744.

[80] Cevc, G., Schätzlein, A., Richardsen, H. Ultradeformable lipid vesicles can penetrate the skin and other semi-permeable barriers unfragmented: evidence from double label CLSM experiments and direct size measurements. *Biochim. Biophys. Acta.* 2002;1564:21-30.

[81] Müller, R. H., Radtke, M., Wissing, S. A. Solid lipid nanoparticles (SLN) and nanostructured lipid carriers (NLC) in cosmetic and dermatological preparations. *Adv. Drug Deliv. Rev.* 2002;54 (Suppl. 1):S131-S155.

[82] Müller, R. H., Radtke, M., Wissing, S. A. Nanostructured lipid matrices for improved microencapsulation of drugs. *Int. J. Pharm.* 2002;24:121-128.

[83] Pardeike, J., Hommoss, A., Müller, R. H. Lipid nanoparticles (SLN, NLC) in cosmetic and pharmaceutical dermal products. *Int. J. Pharm.* 2009;366:170-184.

[84] Pardeike, J., Schmidt, C., Volz, I., Müller, R. H. Nanostructured lipid carriers as delivery system for the phospholipase A2 inhibitors PX-18 and PX-13 for dermal application. *Pharmazie* 2011;66:357-361.

[85] Junyaprasert, V. B., Teeranachaideekul, V., Souto, E. B., Boonme, P., Müller, R. H. Q10-loaded NLC versus nanoemulsions: stability, rheology and in vitro skin permeation. *Int. J. Pharm.* 2009;377:207-214.

[86] Marcato, P. D., Caverzan, J., Rossi-Bergmann, B., Pinto, E. F., Machado, D., Silva, R. A., Justo, G. Z., Ferreira, C. V., Durán, N. Nanostructured polymer and lipid carriers for sunscreen: biological effects and skin permeation. *J. Nanosci. Nanotechnol.* 2011;11:1880-1886.

[87] Ravichandran, R. Nanotechnology applications in food and food processing: innovative green approaches, opportunities and uncertainties for global market. *Int. J. Green Nanotechnol: Phy. Chem.* 2010;1:P72-P96.

[88] Ankola, D. D., Viswanad, B., Bhardwaj, V., Ramarao, P., Kumar, M. N. Development of potent oral nanoparticulate formulation of coenzyme Q10 for treatment of hypertension: can the simple nutritional supplements be used as first line therapeutic agents for prophylaxis/therapy? *Eur. J. Pharm. Biopharm.* 2007;67:361-369.

Newer Targets and Sources for Nutraceutical Development

In: Nutrigenomics
Editor: M. T. Ravi Subbiah

ISBN: 978-1-62417-051-5

Chapter X

Membrane Targets in Nutraceutical Development

William V. Everson[*]
Department of Pediatrics and Physiology
University of Kentucky College of Medicine
Lexington, Kentucky, US

Abstract

Membranes play critical roles in the development of nutraceuticals. In the past two decades, the simple fluid mosaic model of membranes has expanded into a highly sophisticated model. Membranes are now known to consist of multiple domains in a highly dynamic organization in steady state with a complex system of endomembranes and proteins that connect the cell surfaces with organelles. The focus of this chapter is on the whole cell physiological homeostasis of membrane domains known as lipid rafts, and the subset of lipid rafts termed caveolae. These domains in the surface membrane, and similar domains rich in cholesterol and sphingolipid within endomembranes throughout the cell, are organized into balanced sets to maintain the topology of the membranes as required to conserve the primary function as limiting barriers separating both the cell interior from the exterior and delimiting organelles. Understanding the whole cell organization of lipid raft endomembranes as a whole cell network is critical to development of rational approaches for nutraceutical intervention in the pathophysiology of disease.

Keywords: caveolae, lipid rafts, membrane domain, annexins, caveolins, trafficking, targeted delivery, endosome, compartmentation, neutraceutical development

[*] Corresponding author: William V. Everson, Ph.D. Research Associate Professor, Department of Pediatrics and Physiology, MDS Building Room 446, 725 Rose St. Lexington, KY 40536-0082, 859-323-2852, fax 859-257-2120, E-mail: everson@uky.edu.

Introduction

In recent years, systems biology approaches have greatly expanded our knowledge of the structure and organization of cells, tissues and organelles. Advances in the last two decades have changed the way cell biologists think about membranes, as the organization and interaction between surface lipid rafts and endomembranes has emerged. This organelle is comprised of surface domains in steady state equilibrium with endomembranes connecting to all of the organelles in the cell. Lipid rafts, and a subset of them organized and marked by caveolins, are membrane domains enriched in cholesterol and sphingolipids. A wide range and number of functionally distinct protein and protein-lipid complexes are organized in lipid rafts and caveolae. These include signaling complexes, transport complexes (including those involved in calcium homeostasis), proteolytic complexes, proteins involved in cell-cell interaction and communication, and membrane domains involved in the activation of hormone binding, signaling, uptake and targeting. The associated endomembranes are involved in propagation of signals to specific intracellular targets, hormone uptake and delivery, nutrient uptake and transport to sites of catabolism or metabolism and many other processes in the cell.

For any nutraceutical to penetrate a cell and reach its target destination, the nutraceutical first interacts with a macromolecule (e.g., a protein receptor) localized in a specific membrane domain at the cell surface, then the signal propagates or the nutraceutical is transported in a membrane associated fashion to reach a target intracellular destination (Figure 1). Signal transduction and metabolism span across multiple compartments in the cell, requiring coordinated activation of events at multiple sites in conjunction with the activity of trafficking pathways. The purpose of this chapter is to provide a whole cell conceptual model of this network. Signal transduction involves the activation of signals at the surface that have move to the intracellular compartment(s) where the physiological targets are located. These often involve a coordinated activation of second messengers and trafficking pathways that change the physiological homeostasis of the lipid raft endomembrane system.

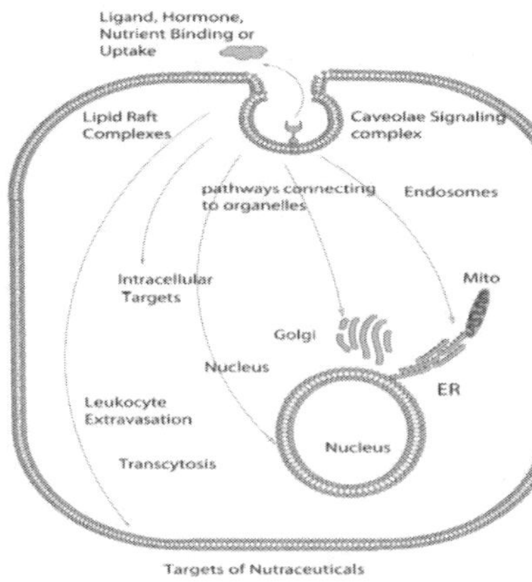

Figure 1. Targets of nutraceutical development involving lipid rafts. Approaches to nutraceutical development depend on the site(s) of action and location within the cell. The lipid raft endomembrane comprises a diverse range of pathways localized to some of the many potential targets shown.

The purpose of this review is to summarize some of the key regulatory elements of the whole cell lipid raft domain system as targets for nutraceutical intervention.

1. The Complexity of Membrane Organization

Systems biology approaches using proteomics and lipidomics have resulted in major advances in our understanding of many cells and diseases that take into account the real world complexity of living systems. This is true as well in the field of membrane biology. A clear illustration of the complexity of membranes was shown by the application of quantitative proteomics and lipidomics applied to the synaptic vesicle [1]. Synaptic vesicles are described as simple trafficking organelles bounded by a spherical phospholipid bilayer membrane whose function is to deliver neurotransmitters to the cell surface upon activation of specific signaling pathways. A careful quantitative study revealed an intriguing model of the synaptic vesicle which gives a clear picture of how complicated this very simple organelle really is [1]:

- The area of helices spanning the membrane taken up by integral membrane proteins and receptors occupied roughly 18% of the volume of the outer membrane and 25% of the inner membrane surface.
- The entire surface was coated by vesicle associated proteins involved in targeting and fusion, including multiple members of the SNARE and rab families.
- There were roughly 7000 phospholipid molecules which would take up approximately 50% of the outer and inner surfaces and 5600 cholesterol molecules per vesicle, making the ration of phospholipid to cholesterol 1: 0.8, with very little sphingomyelin present.
- Modeling the topology of the integral membrane protein helices associated with phospholipids bound to these helices showed that the synaptic vesicle contained no regions with free phospholipids and thus, had no classic 'fluid mosaic' membrane regions in the entire surface.

These non-simple membranes interact with a larger membrane based network involved in the activation of key processes in the nervous system involved in neurotransmission, neuroplasticity, and formation of membrane-membrane connections underlying the central functions of the brain involving thought and memory.

2. Lipid Rafts and Caveolae Are Specialized Domains in Membranes throughout the Cell

2.A. Lipid Rafts Are Found in Apical, Lateral and Basal Surfaces and within Multiple Organelles throughout the Cell Interior

The lipid raft field emerged in 1992 with the cloning of caveolin-1 [2, 3], which was localized to the structures identified in EM micrographs that were termed 'caveolae', 60-80 nm pits at the membrane surface [3]. The first studies showed a caveolae 'proteome that was rich in signal transduction molecules in both endothelial cells [4] and smooth muscle [5], initiating the 'caveolae signaling hypothesis'. This has expanded to include hundreds of signaling pathways that intersect with this whole cell network [6] involving both caveolae and lipid raft domains at apical, lateral and basal surfaces [7]. The field of caveolae and lipid rafts is still expanding, with the identification of new protein families involved in caveolin biogenesis and interaction with endomembrane such as the cavins [8-12].

Caveolae were the first type of lipid raft identified. Ultrastructural work on caveolae goes back to early work by Palade [13], who also described the whole cell organization of caveolae and the dynamic tubular-vesicular structure that comprises a portion of the whole cell network of these interconnected, communicating, lipid raft domains [13]. Caveolae organize a wide range of protein and lipid complexes into distinct functional domains, involved in a diverse but interconnected number of pathways, including the organization of production of nitric oxide by endothelial nitric oxide synthase (eNOS) [14] , calcium influx [15] and homeostasis [15-17], reactive oxygen species produced by NADPH oxidases [18] and many dynamic endocytic and exocytic pathways (reviewed in [7]. Endosomes derived from and communicating with surface caveolae and lipid rafts, play many diverse roles, including transport of signaling complexes within the cell as part of signal propagation [19]. 'Endothelial cell dysfunction' is a broad term describing pathophysiology involved in many diseases. This highlights the central role of caveolae and the associated endomembrane system in the pathophysiology of disease [20].

2.B. Proteins Associated with Caveolae and Lipid Rafts Confer Stability and Serve a Major Scaffolding Role

Many proteins which interact with lipid rafts are responsible for the organization and stabilization of caveolae and lipid rafts. The annexins are a family proteins present in most cell types [17, 21]. Annexin II, unique among the annexins, forms a heterotetramer with an 11 kDa protein in the s100 family now classified as S100A10 and thus can form a bridge between vesicles and actin filaments, and between vesicles and the fusion domain in the plasma membrane [22, 23]. The abundance of annexin in caveolae and lipid rafts [24] highlights the key role annexin 2 plays in the homeostasis of the caveolae-lipid raft endomembrane system. In addition, annexin II may play a role in caveolae stability by providing a linkage between lipid rafts and the actin cytoskeletal meshwork [17, 22, 23]. Intriguingly, annexin II has been identified in caveolae or lipid rafts in most studies using subcellular fractionation to isolate a purified buoyant membrane, lipid raft or caveolae fraction [4, 5] including cells which lack caveolin-1 and thus, only contain lipid rafts [24]. Several studies have shown that annexin II interacts with membranes in either a calcium or cholesterol dependent fashion [25, 26]. The mechanistic basis on which annexin II shifts between a calcium-dependent interaction with phospholipids to a cholesterol dependent, calcium-independent interaction with lipid rafts has not yet been defined mechanistically, but it appears that both phosphorylation and acylation are involved.

Proteins associated with the organization of caveolin are regulated by enzymes which cause covalent modifications. Several proteins associated with lipid rafts are acylated, including caveolin 1 [27-32]. Annexin A2 is also an acylated protein. An early study showed that acylation of annexin was triggered by tyrosine phosphorylation [33], which altered the calcium dependence of membrane interaction. This most basic form representing the tyrosine phosphorylated form of annexin 2 associated with the calcium, independent, Tx-100 insoluble fraction found in fetal membranes [34]. Long chain fatty acids alter the intracellular distribution and caveolae association of both annexin A2 and caveolin-1 [35]. Bromostearate inhibition of acyltransferases was found associated to a change in acylation of caveolin-1 but not annexin A2. Annexin A2 contains several 'reactive cysteines', including one that is a target of oxidation and glutathiolation [36, 37]. The regulation of acylation is reversible in many systems, but the control of acylation and deacylation and the role this plays in caveolae and lipid raft organization and homeostasis is still a field in its infancy. Many proteins found in caveolae and lipid rafts are kinases actively involved in signaling [4, 7] [6]. Caveolae and lipid rafts contain protein interaction domains [31, 32], including the widely recognized scaffolding domain [7] or the presence of domains which allow formation of stable protein complexes that associate with caveolae in large, organized, multi-protein structures. Actin has been found associated with caveolae and lipid rafts in subcellular fractions [4, 5, 31] , and in studies employing electron microscopy, where it has been shown to form a plexus that may confer added stability or allow formation of larger aggregates of lipid rafts, which are linked to dynamic tubular vesicular structures and enzymatic activities critical to the function of macrophages and neutrophils which contain active and highly organized lipid raft tubules and vesicles [38-42].

3. Lipid Raft Associated Complexes on the Cell Surface Define Novel Organization and Specialized Functions

The term 'matricellular' proteins was coined by Paul Bornstein to describe proteins that are organized in multiprotein complexes on the cell surface that control the interactions between cells and the extracellular matrix [43, 44]. This concept has new application to the lipid raft field, as caveolae and lipid rafts appear to organize functional domains for activities localized to the outer leaflet, as well as the well characterized organization of a large number of complexes localized to the inner leaflet. While the role of caveolin and the lipid raft domains themselves have been well described with respect to organization of receptor complexes and signaling complexes, the surface organization has not been well characterized. One of the early functions identified in the annexins, was their ability to be secreted, although they lack classical secretory sequences. In fact, Annexin A2 in its heterotetrameric form is the surface receptor for tissue plasminogen activator [24]. The matricellular concept applies to lipid rafts, in addition to the recognized role of caveolae and lipid rafts in the organization of cytofacial membrane domains. Annexin 2 on the surface has been found in association with a large number of proteases which have been associated with annexin 2 as either a 'receptor' or a ligand, including factor Xa [11], PCSK9 [45, 46], MMP-7 [47], cathepsin B[48], and uPA [49]. It is beginning to emerge that these are regulated both on the surface and within an endomembrane system, as has been described in the activation and control of MMP-2 [50-52]. This links surface matricellular clusters with this dynamic endomembrane lipid raft system. The organization of proteolytic enzymes within this whole cell network in cardiac and smooth muscle has recently been described [16, 51]. The existence of multiple functional networks linked to distinct lipid raft or caveolae domains at the surface, has intriguing implications for defining the acute, interacting regulation of multiple signaling systems.

And thus, for elucidating the basis for pathophysiology as the disruption of the organization of one endomembrane linked system causes an overall imbalance in homeostasis of these interacting lipid raft endomembrane systems. Some very simple concepts of topology and 'conservation of surface area' requirements need to be considered in the whole cell steady state dynamic equilibrium of these systems. CD36 is a scavenger receptor localized to caveolae and lipid raft domains in multiple cell types [53]. CD36, like many receptors found associated with caveolae and lipid rafts, functions both within the domains at the surface and in associated endomembranes. In macrophages, CD36 serves to bind modified lipids at the surface, but also acts within the lipid raft endomembrane system to regulate Toll receptor activation in macrophages [54]. In muscle, caveolin localized CD36 plays a critical role in increasing the uptake of long chain fatty acids and in delivery of fat for metabolism within the caveolin-3 marked caveolae and endomembrane system that links to mitochondria for fatty acid oxidation [55]. In adipose tissue, which has no caveolin-3, but contains abundant caveolin1 and caveolin 2, CD36 regulates fatty acid uptake that is specialized within this endomembrane system for triglyceride accumulation [55-57]. The distinct roles of CD36 in different pathways and cell types is dependent on both the localization and the abundance of distinct proteins that control the homeostasis, such as the caveolins, which differ in abundance and isotype in muscle, fat and macrophages. CD44 is another example of a bifunctional receptor localized to lipid rafts. At the surface, CD44 organizes into large clusters that serve to anchor the cell to the extracellular matrix, or conversely, translocates ligands within the emdomembrane system to propagate signaling molecules which are processed and translocated to distinct intracellular sites, including the nucleus [58]. Recent work in mice deficient in a protein termed cavin uncovered a novel protein involved in the biogenesis of caveolae [11]. Deletion of cavin 1 caused a loss of caveolae resulting in dyslipidemia and glucose intolerance [59]. In smooth muscle, enzymes involved in carbohydrate metabolism have been localized within a caveolin-1 linked endomembrane network [60, 61]. Cardiac muscle exhibits both caveolin-1 and caveolin-3 endomembrane networks which are organized into 2 distinct networks. The annexins also diffentially distributed in these two tissues and to these two networks. These two proteins likely play a critical role in the control of the metabolism and catabolism of fatty acids and carbohydrates [35, 62, 63]. The existence of multiple lipid raft networks which organize distinct metabolic and catabolic pathways in cells, and the interplay with distinct signaling networks offers some intriguing prospects for advancing our understanding of how metabolic and signal transduction networks are organized and coordinately regulated. Defining these networks and how they are disrupted in the pathophysiology of disease states such as obesity and diabetes represents an exciting frontier and opens up new approaches to nutraceutical development.

4. Whole Cell Homeostasis of Lipid Rafts, Caveolae and Associated Intracellular Pathways

Ultrastructural studies have revealed that both the barrier function of the endothelium and transcytosis of macromolecules and even leukocytes across this cell layer involve caveolae at both apical and basal surfaces and an extensive set of small lipid raft vesicles connected in an extensive endomembrane system in the cell [64-69]. This is represented in Figure 2. The VVO is enriched in regions adjacent to lateral membranes and represents as much as 18% of the cytosol volume [70-72] regulating transcytosis and leukocyte extravasation [70-73]. Annexin II-p11, caveolin-1 and cholesterol accumulate together at lateral membranes as endothelial cells become confluent, revealing another significant function of caveolae in maintaining the barrier function. Pecam/CD31 localization in caveolae at apical and lateral surfaces, and in endomembrane tubules and vesicles, has been described and linked to cell-cell adhesion and control of leukocyte extravasation [20, 74].

Several reviews described the whole cell structure and organization of caveolae in endothelial cells, and how this network is related to pathophysiology of disease [20, 73, 75].

Caveolae and lipid rafts are involved in a wide range of dynamic trafficking pathways, as shown in Figure 3. These pathways connect numerous sites within the cell to each other and to distinct regions of the apical, lateral, and basal surfaces. A large number of studies have examined and elucidated pathways involved in signal propagation of nutrient uptake but few studies consider the number and intricate organization and interconnection between these numerous pathways.

The complexity of this set of pathways is illustrated in a recent report which used different ligands that can be transported to distinguish between five endocytic pathways, only one of which is clathrin dependent [59]. Fluorescent lipids and reporters have been used to show the rapid trafficking and identify many of the intracellular sites and connections that comprise this whole cell set of balanced pathways [76-79]. Another concept that is emerging is that the ratio between proteins can change the homeostasis of this system, though little is known about the regulation of expression of the specific proteins or different families of proteins in the overall regulation of this dynamic system.

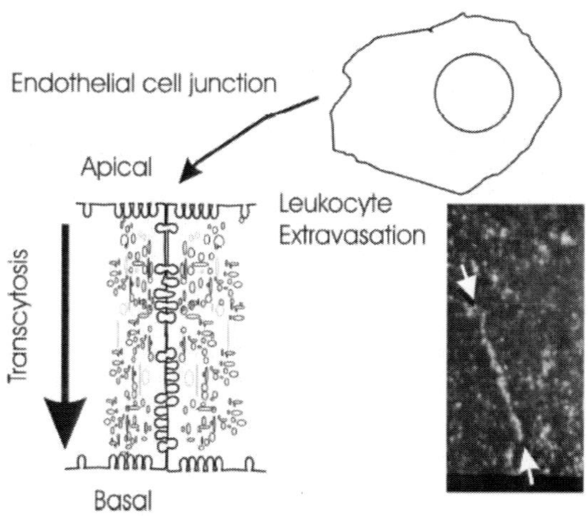

Figure 2. Endothelial transcytosis and leukocyte extravasation. Endothelial cells contain an extensive set of lipid raft domains and associated membrane tubules and vesicles at the lateral domains in confluent cells and capillaries. A cartoon of these structures is shown on the left, and a portion of a confocal slice from confluent human umbilical vein endothelial cells, immunostained with antibody against caveolin 1 is shown at the right. The arrowheads show a region where two endothelial cells are connected where caveolin is localized and highly enriched.

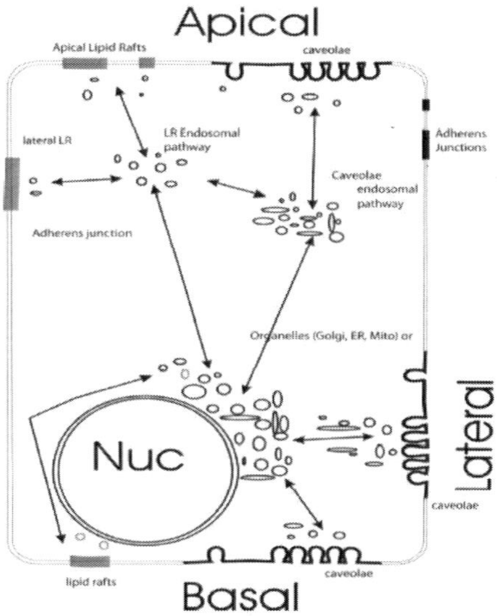

Abbreviations: LR. Lipid rafts. ER. Endoplasmic reticulum. Mito. Mitochondria. Nuc. Nucleus.

Figure 3. Whole cell distribution of caveolae and lipid raft domains. Both caveolae and lipid rafts have been found in apical, lateral and basal membranes, and contained within the many endomembranes including organelles in the cell. These are in a dynamic steady state homeostasis.

It is important to note that even a small amount of 'over'expression of caveolin-1 can cause a change in the dynamics, resulting in 'spillover' of excess caveolin-1 into the lysosomal degradative pathway [80]. Beyond the level of the individual cell, studies examining caveolin-1 and annexin 2 reveal that there are multi-cellular interactions between networks comprising larger structures spanning across multiple cells that form structures important for key tissue properties. For example, annexin 2 plays a critical role along with several other proteins in the organization of signaling complexes and multicellular structures at the surface of cells involved in tubule formation in epithelial tubulogenesis in the kidney and endothelial vasculogenesis in the circulation [81, 82]. Many pathogens exploit lipid raft endomembrane systems to invade cells as part of their life cycle. A large number of bacteria, viruses, and parasites have been found to exploit either lipid rafts or the lipid raft endomembrane system in their takeover of host cells and replicative life cycle, including enterohermorrhagic E. Coli [83, 84], Toxoplasma gondii [85-87] Mycobacteria [88, 89] and Leshmenia [90, 91] alter the lipid raft components during their host cell interaction.

5. Caveolae, Lipid Rafts, and Nutraceutical Development

5.A. Whole Cell Homeostasis Defines Novel Targets for Modulation of ROS and RNS Activity

The regulation of endothelial nitric oxide synthase (eNOS) by caveolin and localization and activation in caveolae has been intensively studied. This regulation involves the whole cell caveolae/lipid raft system, in several key aspects, including the homeostasis of calcium, which is critical for the activation of eNOS [15-17], and the organization of signaling complexes at the cell surface and translocation of the enzyme within the cell to its sites of action [7, 53, 92]. Less well characterized are the pathways and regulation of intracellular assembly and trafficking underlying the control of activity of NADPH oxidases (Nox's). Several Nox's have been localized to caveolae or lipid rafts and translocate within the lipid raft endomembrane system to reach their sites of action. An elegant study in 2004 demonstrated that the activities of two different Nox's, Nox 1 and Nox 4, were localized and activated within two distinct cell surface lipid raft domains. Nox1 localizes to apical caveolae in vascular smooth muscle cells, while Nox4 in the same cells localized to focal adhesions [93]. Reactive oxygen species produced by NADPH oxidases regulate a large number of pathways in the cell [94, 95] and many of these pathways contain enzymes and structural proteins and lipids acting and translocating within the lipid raft endomembrane system. In addition to ROS production by Nox's, ROS production by mitochondria involves lipid rafts. The outer mitochondrial membrane is itself rich in cholesterol, and exhibits the same dynamic properties of other lipid raft endomembranes in many cell types. A recent study has shown that caveolin-1 deficiency has an effect on the outer mitochondrial membrane which leads to mitochondrial dysfunction and elevated ROS production [96] .This change in mitochondrial function was linked impaired fat metabolism and the requirement for glucose in the liver of caveolin-1 null mice [96].

5.B. Modulation of Lipid Raft Homeostasis in Prostate Cancer: Targeting Oxidative Stress to Restore Normal Regulation

The development of prostate cancer and maturation into aggressive forms is accompanied by a progressive loss of annexins, particularly annexin 2, [73, 97] and dysregulation of caveolin-1. Pro-oxidant lipids and cytokines are implicated epidemiologically in prostate cancer development and progression, whereas antioxidants nutrients and vitamins lessen risk and decrease growth or reverse hormone insensitivity in cultured prostate cancer cells [66, 98-101]. Annexin II is a protein sensitive to physiologic oxidation [36]; high/ chronic oxidative stress triggers its proteasomal degradation [102, 103]. The anti-oxidant amino acid taurine binds specifically to annexin II and protects annexin II cysteines from oxidation by NEM in brush border membranes [104]. These studies suggest that antioxidant nutraceuticals that target restoration of annexin 2 homeostasis may be effective agents in helping restore the regulation of multiple growth pathways regulated within the lipid raft endomembrane system.

5.C. Nutraceutical Targeting of Vascular Dysfunction, Endothelial Injury/ Repair and Angiogenesis

Endothelial NOS is a key enzyme regulating vasodilation and angiogenesis which localizes in caveolae [28] where it binds to caveolin through the caveolin scaffolding domain [105, 106]. Caveolin-1 binding to eNOS inhibits the enzyme, and activation of eNOS requires release of caveolin-1, followed by immediate internalization in a lipid raft pathway, modulated by one of the mediators of this homeostasis, e.g., annexin 2. Annexin II in conjunction with p11 has been associated with endocytosis and exocytosis for over two decades [21]. Annexin II and p11 have been shown to associate with the CD44 receptor in its endomembrane trafficking, as discussed above. Annexin 2 and p11 likely play a similar role in the trafficking of eNOS in endothelial cells. Disruption of cholesterol caused both CD44 and annexin II to dissociate and disperse uniformly throughout the membrane, disruption of cholesterol causes both caveolin and eNOS to disperse as well. The activity of eNOS is very carefully regulated within the lipid raft endomembrane system, and the acute sensitivity of some of the key proteins regulating this homeostasis is a clear target for nutraceuticals with antioxidant and lipophillic properties. The phytonutrient resveratol activates eNOS activity within this endomembrane system [107]. Estrogen and the phytoestrogens are clear regulators of both eNOS activity and localization [108, 109]. Epicatechin gallate effects may also be mediated in part by their effects on lipid raft homeostasis, to preserve eNOS activity in the context of Oxidized LDL induced vascular dysfunction [110].

5.D.

Angiogenesis involves both an NO dependent activation of endothelial cell proliferation and multiple processes at the cell surface to allow e endothelial cells to migrate, invade the extracellular matrix, and form tubules. This involves many membrane dependent changes in the cell regulated by lipid raft homeostasis. The importance of this activity in angiogenesis has recently been shown in studies using gene targeted mice. Endothelial cells from these mice exhibit impaired angiogenesis in several angiogenic assays [97, 103]. Annexin II translocation to the cell surface is elevated in response to ‘stress’ including oxidative stress, and the src kinase phosphorylation site, tyr 23, has been implicated in this translocation event [103]. Homocysteinylation of annexin II occurs on a specific residue in the domain that interacts with p11, interferes with tPA binding, and targets annexin II for proteasomal degradation. Recently, in a series of elegant studies, annexin 2 has been shown to play a critical role in tubulogenesis [81, 82].

Conclusion

The many functions regulated within the whole cell membrane system comprised of functional caveolae/lipid raft domains in dynamic steady state in the cell represent novel targets for nutraceutical intervention. Lipid raft and caveolae membrane domains are in a tightly regulated physiologic homeostasis involving rapid and complex trafficking throughout an endomembrane system spanning the whole cell. Aqueous nutraceuticals need to cross membranes through complexes organized within membrane domains, or alternatively, they are taken up and delivered through one or more of the systems comprising the membrane linked endocytic pathways described in this chapter, in order to reach their intracellular targets. Lipophillic compounds interact directly with these organized structures or macromolecules within them with multiple effects, causing changes in the organization of receptor, signaling and surface complexes, changes in steady state distribution and the structures of individual domains. Understanding the organization and regulation of physiological homeostasis within these dynamic, interacting systems, and the imbalances in them that constitute the pathophysiology of disease, is critical to a rational approach to development of nutraceuticals.

References

[1] Takamori, S., Holt, M., Stenius, K., Lemke, E. A., Gronborg, M., Riedel, D., Urlaub, H., Schenck, S., Brugger, B., Ringler, P., Muller, S. A., Rammner, B., Grater, F., Hub, J. S., De Groot, B. L., Mieskes, G., Moriyama, Y., Klingauf, J., Grubmuller, H., Heuser, J., Wieland, F., and Jahn, R. Molecular anatomy of a trafficking organelle. *Cell.* 2006; 127: 831-46.

[2] Glenney, J. R., Jr. The sequence of human caveolin reveals identity with VIP21, a component of transport vesicles. *FEBS Lett.* 1992; 314: 45-8.

[3] Rothberg, K. G., Heuser, J. E., Donzell, W. C., Ying, Y. S., Glenney, J. R., and Anderson, R. G. Caveolin, a protein component of caveolae membrane coats. *Cell.* 1992; 68: 673-82.

[4] Lisanti, M. P., Scherer, P. E., Vidugiriene, J., Tang, Z., Hermanowski-Vosatka, A., Tu, Y. H., Cook, R. F., and Sargiacomo, M. Characterization of caveolin-rich membrane domains isolated from an endothelial-rich source: implications for human disease. *J. Cell Biol.* 1994; 126: 111-126.

[5] Chang., W. J., Ying, Y. S., Rothberg, K. G., Hooper, N. M., Turner, A. J., Gambliel, H. A., De Gunzburg, J., Mumby, S. M., Gilman, A. G., and Anderson, R. G. Purification and characterization of smooth muscle cell caveolae. *J. Cell Biol.* 1994; 126: 127-38.

[6] Anderson, R. G. The caveolae membrane system. *Annu. Rev. Biochem.* 1998; 67: 199-225.

[7] Bastiani, M. and Parton, R. G. Caveolae at a glance. *J. Cell Sci.* 2010; 123: 3831-6.

[8] Briand, N., Le Lay, S., Sessa, W. C., Ferre, P., and Dugail, I. Distinct roles of endothelial and adipocyte caveolin-1 in macrophage infiltration and adipose tissue metabolic activity. *Diabetes.* 2010; 60: 448-53.

[9] Zheng, Y. Z., Boscher, C., Inder, K. R., Fairbank, M. R., Loo, D. R., Hill, M. M., Nabi, I. R., and Foster, L. J. Differential impact of caveolae and caveolin-1 scaffolds on the membrane raft proteome. *Mol. Cell Proteomics.* 2011.

[10] Hansen, C. G. and Nichols, B. J. Exploring the caves: cavins, caveolins and caveolae. *Trends Cell Biol.* 2010; 20: 177-86.

[11] Bhattacharjee, G., Ahamed, J., Pawlinski, R., Liu, C., Mackman, N., Ruf, W., and Edgington, T. S. Factor Xa binding to annexin 2 mediates signal transduction via protease-activated receptor 1. *Circ. Res.* 2008; 102: 457-64.

[12] Bastiani, M., Liu, L., Hill, M. M., Jedrychowski, M. P., Nixon, S. J., Lo, H. P., Abankwa, D., Luetterforst, R., Fernandez-Rojo, M., Breen, M. R., Gygi, S. P., Vinten, J., Walser, P. J., North, K. N., Hancock, J. F., Pilch, P. F., and Parton, R. G. MURC/Cavin-4 and cavin family members form tissue-specific caveolar complexes. *J. Cell Biol.* 2009; 185: 1259-73.

[13] Palade, G. E. Fine structure of blood capillaries. *J. Appl. Physiol.* 1953: 1424-1436.

[14] Rath, G., Dessy, C. and Feron, O. Caveolae, caveolin and control of vascular tone: nitric oxide (NO) and endothelium derived hyperpolarizing factor (EDHF) regulation. *J. Physiol. Pharmacol.* 2009; 60 Suppl. 4: 105-9.

[15] Pani, B. and Singh, B. B. Lipid rafts/caveolae as microdomains of calcium signaling. *Cell Calcium.* 2009; 45: 625-33.

[16] Daniel, E. E., El-Yazbi, A. and Cho, W. J. Caveolae and calcium handling, a review and a hypothesis. *J. Cell. Mol. Med.* 2006; 10: 529-44.

[17] Gerke, V., Creutz, C. E. and Moss, S. E. Annexins: linking Ca2+ signalling to membrane dynamics. *Nat. Rev. Mol. Cell. Biol.* 2005; 6: 449-61.

[18] Shao, D., Segal, A. W. and Dekker, L. V. Lipid rafts determine efficiency of NADPH oxidase activation in neutrophils. *FEBS Lett.* 2003; 550: 101-6.

[19] Gould, G. W. and Lippincott-Schwartz, J. New roles for endosomes: from vesicular carriers to multi-purpose platforms. *Nat. Rev. Mol. Cell. Biol.* 2009; 10: 287-92.

[20] Simionescu, M. Implications of early structural-functional changes in the endothelium for vascular disease. *Arterioscler. Thromb. Vasc. Biol.* 2007; 27: 266-74.

[21] Gerke, V. and Moss, S. E. Annexins: from structure to function. *Physiol. Rev.* 2002; 82: 331-71.

[22] Hayes, M. J., Shao, D. M., Grieve, A., Levine, T., Bailly, M., and Moss, S. E. Annexin A2 at the interface between F-actin and membranes enriched in phosphatidylinositol 4,5,-bisphosphate. *Biochim. Biophys. Acta.* 2009; 1793: 1086-95.

[23] Hayes, M. J., Shao, D., Bailly, M., and Moss, S. E. Regulation of actin dynamics by annexin 2. *Embo. J.* 2006; 25: 1816-26.

[24] Harder, T. and Gerke, V. The annexin II2p11(2) complex is the major protein component of the triton X-100-insoluble low-density fraction prepared from MDCK cells in the presence of Ca2+. *Biochim. Biophys. Acta.* 1994; 1223: 375-82.

[25] Zeuschner, D., Stoorvogel, W. and Gerke, V. Association of annexin 2 with recycling endosomes requires either calcium- or cholesterol-stabilized membrane domains. *Eur. J. Cell. Biol.* 2001; 80: 499-507.

[26] Ayala-Sanmartin, J. Cholesterol enhances phospholipid binding and aggregation of annexins by their core domain. *Biochem. Biophys. Res. Commun.* 2001; 283: 72-9.

[27] McCabe, J. B. and Berthiaume, L. G. N-terminal protein acylation confers localization to cholesterol, sphingolipid-enriched membranes but not to lipid rafts/caveolae. *Mol. Biol. Cell.* 2001; 12: 3601-17.

[28] Shaul, P. W., Smart, E. J., Robinson, L. J., German, Z., Yuhanna, I. S., Ying, Y., Anderson, R. G., and Michel, T. Acylation targets emdothelial nitric-oxide synthase to plasmalemmal caveolae. *J. Biol. Chem.* 1996; 271: 6518-6522.

[29] Uittenbogaard, A. and Smart, E. J. Palmitoylation of caveolin-1 is required for cholesterol binding, chaperone complex formation, and rapid transport of cholesterol to caveolae. *J. Biol. Chem.* 2000; 275: 25595-25599.

[30] Uittenbogaard, A., Everson, W. V., Matveev, S. V., and Smart, E. J. Cholesteryl ester is transported from caveolae to internal membranes as part of a caveolin-annexin II lipid-protein complex. *J. Biol. Chem.* 2002; 277: 4925-4931.

[31] Smart, E. J., Graf, G. A., McNiven, M. A., Sessa, W. C., Engelman, J. A., Scherer, P. E., Okamoto, T., and Lisanti, M. P. Caveolins, liquid-ordered domains, and signal transduction. *Mol. Cell. Biol.* 1999; 19: 7289-304.

[32] Fernandez, I., Ying, Y., Albanesi, J., and Anderson, R. G. Mechanism of caveolin filament assembly. *Proc. Natl. Acad. Sci. US* 2002; 99: 11193-11198.
[33] Soric, J. and Gordon, J. A. The 36-kilodalton substrate of pp60v-src is myristylated in a transformation-sensitive manner. *Science*. 1985; 230: 563-6.
[34] Myatt, L., Hirth, J. and Everson, W. V. Changes in annexin (lipocortin) content in human amnion and chorion at parturition. *J. Cell. Biochem*. 1992; 50: 363-73.
[35] Zhao, H. and Hardy, R. W. Long-chain saturated fatty acids induce annexin II translocation to detergent-resistant membranes. *Biochem. J*. 2004; 381: 463-9.
[36] Sullivan, D. M., Levine, R. L. and Finkel, T. Detection and affinity purification of oxidant-sensitive proteins using biotinylated glutathione ethyl ester. *Methods Enzymol.* 2002; 353: 101-13.
[37] Sullivan, D. M., Wehr, N. B., Fergusson, M. M., Levine, R. L., and Finkel, T. Identification of oxidant-sensitive proteins: TNF-alpha induces protein glutathiolation. *Biochemistry*. 2000; 39: 11121-8.
[38] Robinson, J. M., Kobayashi, T., Seguchi, H., and Takizawa, T. Evaluation of neutrophil structure and function by electron microscopy: cytochemical studies. *J. Immunol. Methods*. 1999; 232: 169-78.
[39] Kobayashi, T., Robinson, J. M. and Seguchi, H. Identification of intracellular sites of superoxide production in stimulated neutrophils. *J. Cell Sci*. 1998; 111 (Pt 1): 81-91.
[40] Kobayashi, T. and Robinson, J. M. A novel intracellular compartment with unusual secretory properties in human neutrophils. *J. Cell Biol*. 1991; 113: 743-56.
[41] Parton, R. G. and Richards, A. A. Lipid rafts and caveolae as portals for endocytosis: new insights and common mechanisms. *Traffic*. 2003; 4: 724-738.
[42] Mundy, D. I., Machleidt, T., Ying, Y. S., Anderson, R. G., and Bloom, G. S. Dual control of caveolar membrane traffic by microtubules and the actin cytoskeleton. *J. Cell Sci*. 2002; 115: 4327-39.
[43] Bornstein, P. and Sage, E. H. Matricellular proteins: extracellular modulators of cell function. *Curr. Opin. Cell. Biol*. 2002; 14: 608-16.
[44] Sage, E. H. and Bornstein, P. Extracellular proteins that modulate cell-matrix interactions. SPARC, tenascin, and thrombospondin. *J. Biol. Chem*. 1991; 266: 14831-4.
[45] Seidah, N. G. PCSK9 as a therapeutic target of dyslipidemia. *Expert Opin. Ther. Targets*. 2009; 13: 19-28.
[46] Mayer, G., Poirier, S. and Seidah, N. G. Annexin A2 is a C-terminal PCSK9-binding protein that regulates endogenous low density lipoprotein receptor levels. *J. Biol. Chem.* 2008; 283: 31791-801.
[47] Tsunezumi, J., Yamamoto, K., Higashi, S., and Miyazak,i K. Matrilysin (matrix metalloprotease-7) cleaves membrane-bound annexin II and enhances binding of tissue-type plasminogen activator to cancer cell surfaces. *FEBS J*. 2008; 275: 4810-23.
[48] Roshy, S., Sloane, B. F. and Moin, K. Pericellular cathepsin B and malignant progression. *Cancer Metastasis Rev*. 2003; 22: 271-86.
[49] Zhang, Y., Zhou, Z. H., Bugge, T. H., and Wahl, L. M. Urokinase-type plasminogen activator stimulation of monocyte matrix metalloproteinase-1 production is mediated by plasmin-dependent signaling through annexin A2 and inhibited by inactive plasmin. *J. Immunol*. 2007; 179: 3297-304.

[50] Ali, M. A. and Schulz, R. Activation of MMP-2 as a key event in oxidative stress injury to the heart. *Front. Biosci.* 2009; 14: 699-716.

[51] Cho, W. J., Chow, A. K., Schulz, R., and Daniel, E. E. Matrix metalloproteinase-2, caveolins, focal adhesion kinase and c-Kit in cells of the mouse myocardium. *J. Cell Mol. Med.* 2007; 11: 1069-86.

[52] Kandasamy, A. D., Chow, A. K., Ali, M. A., and Schulz, R. Matrix metalloproteinase-2 and myocardial oxidative stress injury: beyond the matrix. *Cardiovasc. Res.* 2009.

[53] Silverstein, R. L. and Febbraio, M. CD36, a scavenger receptor involved in immunity, metabolism, angiogenesis, and behavior. *Sci. Signal.* 2009; 2: re3.

[54] Stewart, C. R., Stuart, L. M., Wilkinson, K., van Gils, J. M., Deng, J., Halle, A., Rayner, K. J., Boyer, L., Zhong, R., Frazier, W. A., Lacy-Hulbert, A., El Khoury, J., Golenbock, D. T., and Moore, K. J. CD36 ligands promote sterile inflammation through assembly of a Toll-like receptor 4 and 6 heterodimer. *Nat. Immunol.* 2010; 11: 155-61.

[55] Holloway, G. P. Mitochondrial function and dysfunction in exercise and insulin resistance. *Appl. Physiol. Nutr. Metab.* 2009; 34: 440-6.

[56] Holloway, G. P., Jain, S. S., Bezaire, V., Han, X. X., Glatz, J. F., Luiken, J. J., Harper, M. E., and Bonen, A. FAT/CD36-null mice reveal that mitochondrial FAT/CD36 is required to upregulate mitochondrial fatty acid oxidation in contracting muscle. *Am. J. Physiol. Regul. Integr. Comp. Physiol.* 2009; 297: R960-7.

[57] Aguer, C., Mercier, J., Man, C. Y., Metz, L., Bordenave, S., Lambert, K., Jean, E., Lantier, L., Bounoua, L., Brun, J. F., Raynaud de Mauverger, E., Andreelli, F., Foretz, M., and Kitzmann, M. Intramyocellular lipid accumulation is associated with permanent relocation ex vivo and in vitro of fatty acid translocase (FAT)/CD36 in obese patients. *Diabetologia.* 2011; 53: 1151-63.

[58] Oliferenko, S., Paiha, K., Harder, T., Gerke, V., Schwarzler, C., Schwarz, H., Beug, H., Gunthert, U., and Huber, L. A. Analysis of CD44-containing lipid rafts: Recruitment of annexin II and stabilization by the actin cytoskeleton. *J. Cell Biol.* 1999; 146: 843-54.

[59] Simard, J. R., Meshulam, T., Pillai, B. K., Kirber, M. T., Brunaldi, K., Xu, S., Pilch, P. F., and Hamilton, J. A. Caveolins sequester fatty acids on the cytoplasmic leaflet of the plasma membrane, augment triglyceride formation and protect cells from lipotoxicity. *J. Lipid Res.* 2009.

[60] Vallejo, J. and Hardin, C. D. Metabolic organization in vascular smooth muscle: distribution and localization of caveolin-1 and phosphofructokinase. *Am. J. Physiol. Cell Physiol.* 2004; 286: C43-54.

[61] Lloyd, P. G. and Hardin, C. D. Caveolae and the organization of carbohydrate metabolism in vascular smooth muscle. *J. Cell Biochem.* 2001; 82: 399-408.

[62] Zaks, W. J. and Creutz, C. E. Annexin-chromaffin granule membrane interactions: a comparative study of synexin, p32 and p67. *Biochim. Biophys. Acta.* 1990; 1029: 149-60.

[63] Creutz, C. E. cis-Unsaturated fatty acids induce the fusion of chromaffin granules aggregated by synexin. *J. Cell Biol.* 1981; 91: 247-56.

[64] Bruns, R. R. and Palade, G. E. Studies on blood capillaries. II. Transport of ferritin molecules across the wall of muscle capillaries. *J. Cell Biol.* 1968; 37: 277-99.

[65] Predescu, D., Predescu, S., McQuistan, T., and Palade, G. E. Transcytosis of alpha1-acidic glycoprotein in the continuous microvascular endothelium. *Proc. Natl. Acad. Sci. US* 1998; 95: 6175-80.

[66] Predescu, S. A., Predescu, D. N. and Palade, G. E. Plasmalemmal vesicles function as transcytotic carriers for small proteins in the continuous endothelium. *Am. J. Physiol.* 1997; 272: H937-49.

[67] Roberts, W. G. and Palade, G. E. Increased microvascular permeability and endothelial fenestration induced by vascular endothelial growth factor. *J. Cell Sci.* 1995; 108 (Pt 6): 2369-79.

[68] Simionescu, M., Simionescu, N., Santoro, F., and Palade, G. E. Differentiated microdomains of the luminal plasmalemma of murine muscle capillaries: segmental variations in young and old animals. *J. Cell Biol.* 1985; 100: 1396-407.

[69] Stan, R. V., Kubitza, M. and Palade, G. E. PV-1 is a component of the fenestral and stomatal diaphragms in fenestrated endothelia. *Proc. Natl. Acad. Sci. US* 1999; 96: 13203-7.

[70] Feng, D., Nagy, J. A., Dvorak, H. F., and Dvorak, A. M. Ultrastructural studies define soluble macromolecular, particulate, and cellular transendothelial cell pathways in venules, lymphatic vessels, and tumor-associated microvessels in man and animals. *Microsc. Res. Tech.* 2002; 57: 289-326.

[71] Dvorak, A. M., Kohn, S., Morgan, E. S., Fox, P., Nagy, J. A., and Dvorak, H. F. The vesiculo-vacuolar organelle (VVO): a distinct endothelial cell structure that provides a transcellular pathway for macromolecular extravasation. *J. Leukoc. Biol.* 1996; 59: 100-115.

[72] Dvorak, A. M. and Feng, D. The vesiculo-vacuolar organelle (VVO). A new endothelial cell permeability organelle. *J. Histochem. Cytochem.* 2001; 49: 419-432.

[73] Burns, A. R., Smith, C. W. and Walker, D. C. Unique structural features that influence neutrophil emigration into the lung. *Physiol. Rev.* 2003; 83: 309-36.

[74] Feng, D., Nagy, J. A., Pyne, K., Dvorak, H. F., and Dvorak, A. M. Ultrastructural localization of platelet endothelial cell adhesion molecule (PECAM-1, CD31) in vascular endothelium. *J. Histochem. Cytochem.* 2004; 52: 87-101.

[75] Simionescu, M. and Antohe, F. Functional ultrastructure of the vascular endothelium: changes in various pathologies. *Handb. Exp. Pharmacol.* 2006: 41-69.

[76] Schroeder, F., Gallegos, A. M., Atshaves, B. P., Storey, S. M., McIntosh, A. L., Petrescu, A. D., Huang, H., Starodub, O., Chao, H., Yang, H., Frolov, A., and Kier, A. B. Recent advances in membrane microdomains: rafts, caveolae, and intracellular cholesterol trafficking. *Exp. Biol. Med. (Maywood).* 2001; 226: 873-90.

[77] Schroeder, F., Atshaves, B. P., McIntosh, A. L., Gallegos, A. M., Storey, S. M., Parr, R. D., Jefferson, J. R., Ball, J. M., and Kier, A. B. Sterol carrier protein-2: new roles in regulating lipid rafts and signaling. *Biochim. Biophys. Acta.* 2007; 1771: 700-18.

[78] Schroeder, F., Huang, H., McIntosh, A. L., Atshaves, B. P., Martin, G. G., and Kier, A. B. Caveolin, sterol carrier protein-2, membrane cholesterol-rich microdomains and intracellular cholesterol trafficking. *Subcell Biochem.* 2010; 51: 279-318.

[79] Polishchuk, R., Di Pentima, A. and Lippincott-Schwartz, J. Delivery of raft-associated, GPI-anchored proteins to the apical surface of polarized MDCK cells by a transcytotic pathway. *Nat. Cell Biol.* 2004; 6: 297-307.

[80] Parton, R. G. and Howes, M. T. Revisiting caveolin trafficking: the end of the caveosome. *J. Cell Biol.* 2010; 191: 439-41.

[81] Martin-Belmonte, F., Gassama, A., Datta, A., Yu, W., Rescher, U., Gerke, V., and Mostov, K. PTEN-mediated apical segregation of phosphoinositides controls epithelial morphogenesis through Cdc42. *Cell*. 2007; 128: 383-97.

[82] Davis, G. E., Stratman, A. N., Sacharidou, A., and Koh, W. Molecular basis for endothelial lumen formation and tubulogenesis during vasculogenesis and angiogenic sprouting. *Int. Rev. Cell Mol. Biol*. 2010; 288: 101-65.

[83] Tobe, T. Cytoskeleton-modulating effectors of enteropathogenic and enterohemorrhagic Escherichia coli: role of EspL2 in adherence and an alternative pathway for modulating cytoskeleton through Annexin A2 function. *FEBS J*. 2010; 277: 2403-8.

[84] Boettcher, J. P., Kirchner, M., Churin, Y., Kaushansky, A., Pompaiah, M., Thorn, H., Brinkmann, V., Macbeath, G., and Meyer, T. F. Tyrosine-phosphorylated caveolin-1 blocks bacterial uptake by inducing Vav2-RhoA-mediated cytoskeletal rearrangements. *PLoS Biol*. 2010; 8.

[85] Coppens, I. and Joiner, K. A. Host but not parasite cholesterol controls Toxoplasma cell entry by modulating organelle discharge. *Mol. Biol. Cell*. 2003; 14: 3804-20.

[86] Charron, A. J. and Sibley, L. D. Molecular partitioning during host cell penetration by Toxoplasma gondii. *Traffic*. 2004; 5: 855-67.

[87] Sehgal, A., Bettiol, S., Pypaert, M., Wenk, M. R., Kaasch, A., Blader, I. J., Joiner, K. A., and Coppens, I. Peculiarities of host cholesterol transport to the unique intracellular vacuole containing Toxoplasma. *Traffic*. 2005; 6: 1125-41.

[88] Jozefowski, S., Sobota, A. and Kwiatkowska, K. How Mycobacterium tuberculosis subverts host immune responses. *Bioessays*. 2008; 30: 943-54.

[89] Maldonado-Garcia, G., Chico-Ortiz, M., Lopez-Marin, L. M., and Sanchez-Garcia, F. J. High-polarity Mycobacterium avium-derived lipids interact with murine macrophage lipid rafts. *Scand. J. Immunol*. 2004; 60: 463-70.

[90] Rodriguez, N. E., Gaur, U. and Wilson, M. E. Role of caveolae in Leishmania chagasi phagocytosis and intracellular survival in macrophages. *Cell Microbiol*. 2006; 8: 1106-20.

[91] Rodriguez, N. E., Gaur Dixit, U., Allen, L. A., and Wilson, M. E. Stage-specific pathways of Leishmania infantum chagasi entry and phagosome maturation in macrophages. *PLoS One*. 2011; 6: e19000.

[92] Feron, O. and Balligand, J. L. Caveolins and the regulation of endothelial nitric oxide synthase in the heart. *Cardiovasc. Res*. 2006; 69: 788-97.

[93] Hilenski, L. L., Clempus, R. E., Quinn, M. T., Lambeth, J. D., and Griendling, K. K. Distinct subcellular localizations of Nox1 and Nox4 in vascular smooth muscle cells. *Arterioscler. Thromb. Vasc. Biol*. 2004; 24: 677-83.

[94] Brown, D. I. and Griendling, K. K. Nox proteins in signal transduction. *Free Radic. Biol. Med*. 2009; 47: 1239-53.

[95] Al Ghouleh, I., Khoo, N. K., Knaus, U. G., Griendling, K. K., Touyz, R. M., Thannickal, V. J., Barchowsky, A., Nauseef, W. M., Kelley, E. E., Bauer, P. M., Darley-Usmar, V., Shiva, S., Cifuentes-Pagano, E., Freeman, B. A., Gladwin, M. T., and Pagano, P. J. Oxidases and peroxidases in cardiovascular and lung disease: New concepts in reactive oxygen species signaling. *Free Radic. Biol. Med*. 2010.

[96] Bosch, M., Mari, M., Gross, S. P., Fernandez-Checa, J. C., and Pol, A. Mitochondrial cholesterol: a connection between caveolin and metabolism. *Traffic*. 2011.

[97] Chetcuti, A., Margan, S. H., Russell, P., Mann, S., Millar, D. S., Clark, S. J., Rogers, J., Handelsman, D. J., and Dong, Q. Loss of annexin II heavy and light chains in prostate cancer and its precursors. *Cancer Res.* 2001; 61: 6331-4.

[98] Kuller, L. H. Dietary fat and chronic diseases: epidemiologic overview. *J. Am. Diet Assoc.* 1997; 97: S9-15.

[99] Newcomer, L. M., King, I. B., Wicklund, K. G., and Stanford, J. L. The association of fatty acids with prostate cancer risk. *Prostate.* 2001; 47: 262-8.

[100] Shirai, T., Asamoto, M., Takahashi, S., and Imaida, K. Diet and prostate cancer. *Toxicology.* 2002; 181-182: 89-94.

[101] Stemmermann, G. N., Nomura, A. M. and Heilbrun, L. K. Cancer risk in relation to fat and energy intake among Hawaii Japanese: a prospective study. *Princess Takamatsu Symp.* 1985; 16: 265-74.

[102] Hajjar, K. A. and Jacovina, A. T. Modulation of annexin II by homocysteine: implications for atherothrombosis. *J. Investig. Med.* 1998; 46: 364-9.

[103] Deora, A. B., Kreitzer, G., Jacovina, A. T., and Hajjar, K. A. An annexin 2 phosphorylation switch mediates p11-dependent translocation of annexin 2 to the cell surface. *J. Biol. Chem.* 2004; 279: 43411-8.

[104] Dumaswala, R. and Brown, T. L. Sulphydryl modification inhibits taurine transport in human placental brush border membranes. *Placenta.* 1996; 17: 329-36.

[105] Garcia-Cardena, G., Fan, R., Stern, D. F., Liu, J., and Sessa, W. C. Endothelial nitric oxide synthase is regulated by tyrosine phosphorylation and interacts with caveolin-1. *J. Biol. Chem.* 1996; 271: 27237-40.

[106] Garcia-Cardena, G., Oh, P., Liu, J., Schnitzer, J. E., and Sessa, W. C. Targeting of nitric oxide synthase to endothelial cell caveolae via palmitoylation: implications for nitric oxide signaling. *Proc. Natl. Acad. Sci. US* 1996; 93: 6448-53.

[107] Penumathsa, S. V., Thirunavukkarasu, M., Zhan, L., Maulik, G., Menon, V. P., Bagchi, D., and Maulik, N. Resveratrol enhances GLUT-4 translocation to the caveolar lipid raft fractions through AMPK/Akt/eNOS signalling pathway in diabetic myocardium. *J. Cell Mol. Med.* 2008; 12: 2350-61.

[108] Lang, U., Baker, R. S. and Clark, K. E. Estrogen-induced increases in coronary blood flow are antagonized by inhibitors of nitric oxide synthesis. *Eur. J. Obstet. Gynecol. Reprod. Biol.* 1997; 74: 229-35.

[109] Van Buren, G. A., Yang, D. S. and Clark, K. E. Estrogen-induced uterine vasodilatation is antagonized by L-nitroarginine methyl ester, an inhibitor of nitric oxide synthesis. *Am. J. Obstet. Gynecol.* 1992; 167: 828-33.

[110] Steffen, Y., Schewe, T. and Sies, H. Epicatechin protects endothelial cells against oxidized LDL and maintains NO synthase. *Biochem. Biophys. Res. Commun.* 2005; 331: 1277-83.

In: Nutrigenomics
Editor: M. T. Ravi Subbiah
ISBN: 978-1-62417-051-5

Chapter XI

Lepidium Meyenii (Maca): Excellent Source for Nutrigenomic Exploration and the Development of Nutraceuticals for Fertility and Sexual Health

Gustavo F. Gonzales*
Department of Physiological Sciences, Faculty of Sciences and Philosophy, and Instituto de Investigaciones de la Altura, Universidad Peruana Cayetano Heredia, Lima, Peru

Abstract

Lepidium meyenii (maca) is a Peruvian plant of the Brassicaceae family, which is cultivated and grows exclusively in the Peruvian central Andes between 4000 and 4500 m altitude. Maca is used as a food supplement in the central Andes and also by its medicinal properties described traditionally. Since the 1990´s of the XX century, an increasing interest on products from maca has been observed in many parts of the world. In the last decade exportation of maca from Peru has increased from 1'415,000 USD in 2001 to USD 6'170,000 USD in 2010. Experimental scientific evidences showed that maca has nutritional, energizer, fertility-enhancer properties and acts on sexual dysfunctions, osteoporosis, benign prostatic hyperplasia, memory and learning, and protect skin against ultraviolet radiation. Clinical trials showed efficacy of maca on sexual dysfunctions as well as increasing sperm count and fertility. Maca is a plant with great potential as an adaptogen and appears to be a promising source for nutraceuticals in the prevention of several diseases. Very little nutrigenomic studies have been done so far on maca and its active components, providing an excellent opportunity to explore this plant for the development of potential nutraceuticals for fertility and sexual health.

* Corresponding author: Gustavo F. Gonzales. Department of Biological and Physiological Sciences, Faculty of Sciences and Philosophy, and Instituto de Investigaciones de la Altura, Universidad Peruana Cayetano Heredia, Av. Honorio Delgado 430, Lima 31, Peru. E-mail: Gustavo.gonzales@upch.pe.

Keywords: nutraceutical, maca, lepidum meyenii, fertility, reproduction

Introduction

Lepidium meyenii Walpers (maca) is a Peruvian plant that belongs to the Brassicaceae family (Cruciferae) [1]. Maca mainly grows at an altitude of ≥4,000 m in a habitat of intense cold, extremely intense sunlight and strong winds. Maca has been cultivated in the Peruvian central Andes, in the former Chinchaycocha (Plateau of Bombón); present-day: Carhuamayo, Junin, and Óndores in the Junin Plateau close to Cerro de Pasco [2].

Maca was probably domesticated in San Blas, Junin (present-day: Ondores) some 1,300–2,000 years ago. Maca is used as a food supplement and for its presumed medicinal properties [3]. Peru's native population in the central Andes uses the dry hypocotyls in amounts >20 g/d. There are no reports of adverse reactions after consuming *Lepidium meyenii* in food [4].

Natives from the highlands of Peru recommend that maca be boiled before its consumption because fresh maca may have adverse effects on health [5]. The effects of fresh maca on health have not been scientific assessed yet. Historical aspects and biological properties of maca, gathered from experimental and clinical studies on this species reveal the importance of this plant as nutraceutical food and that maca has been adapted to harsh conditions as observed at high altitudes [2,3,5-7]

Historical Aspects

The first written description about maca (as a root without identification of the botanical or popular name) was published in 1553, in which Cieza de Leon, a chronicler of the Spaniard conquest of Peru noted that in the Peruvian highlands, particularly in the province of Bombón (Chinchaycocha) (Present day: Junin) the natives used certain roots for consumption [6]. The roots, he was referring was maca. Father Cobo [2] was the first to describe the name of maca and its properties in 1653. He stated that this plant grows in the harshest and coldest areas of the province of Chinchaycocha where no other plant for man's sustenance could be grown. Cobo also referred to the use of maca for fertility. In the 18th century, Hipólito Ruiz referred to the fertility-enhancing properties of maca and also its energizer or stimulant effects [7].

Traditionally, after harvesting, maca is dried naturally and can be stored for many years [5]. The dried hypocotyls are as hard as stone (Figure 1). After natural drying maca hypocotyls can be eaten. Before eating, the hypocotyls need to be boiled in water to obtain a soft product which can be consumed as juice, the most frequent form of use [4]. Temperature will affect the availability of several secondary metabolites in maca causing increases in some metabolites and a reduction in others. One of the important constituents in maca is the glucosinolates. These compounds are sensitive to heating [8]. Similar changes to heating have been noted in broccoli and tomato [9,10]. In broccoli, heating decreases the activity of epithiospecifier protein and increases formation of sulforaphane, a derivative of isothiocyanates and glucosinolates. In tomato, after 2, 15, and 30 min of heating at 88 °C, the vitamin C content and antioxidant activity drops significantly. Yet, the content of trans-lycopene per gram of tomato increases [10].

Biology

Maca is characterized both as an aboveground and an underground part. The aboveground part is small and flat in appearance. This seems to be the result of an adaptation process to prevent the impact of strong winds. The principal and the edible part of the plant is a radish-like tuber that constitutes the hypocotyl and the root of the plant. This hypocotyl-root axis is 10–14 cm long and 3–5 cm wide and constitutes the storage organ storing a high content of water. After drying the hypocotyls are dramatically reduced in size to about 2–8 cm in diameter (Figure 1). The average weight of the dried hypocotyls may vary considerably ranging from 7.64 g in hypocotyls from black maca in Yanacancha, Chupaca, to 23.88 g in black maca from Carhuamayo, Junin (both in the Peruvian central Andes).

There are many types of maca that can be characterized by the color of their hypocotyls. In Carhuamayo, Junin, in the Peruvian highlands, 13 colors of maca have been described, ranging from white to black [11]. Recently, it has been demonstrated that different types of maca (according to its color) have different biological properties [12–14].

Chemistry

Primary metabolites correspond to the nutritional component of the hypocotyls, and the secondary metabolites are associated with biological and medicinal properties.

Primary Metabolites

The dried hypocotyls of maca contain approximately 13–16% protein, and are rich in essential amino acids. Fresh hypocotyls contain 80% water and have high amounts of iron and calcium [see 5].

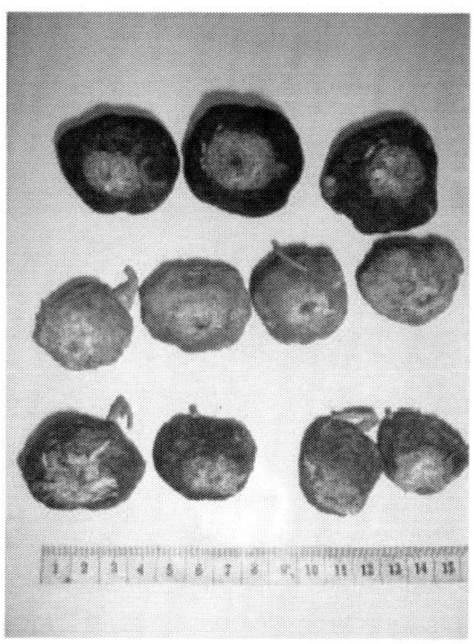

Figure1. Dried hypocotyls of naturally dried black (upper), yellow (middle) and red (bottom) maca.

A more complete description of the composition of dry maca was published by Dini et al.: [15] 10.2% proteins; 59% carbohydrates; 2.2% lipids; and 8.5% of fibre. Free fatty acids are also present in maca, the most abundant being linoleic, palmitic and oleic acids. Saturated fatty acids represent 40.1%, whereas unsaturated fatty acids are present at 52.7%.

Maca contains amino acids (mg/g protein) like leucine (91.0 mg), arginine (99.4 mg), phenylalanine (55.3 mg), lysine (54.3 mg), glycin (68.30 mg), alanine (63.1 mg), valine (79.3 mg), isoleucine (47.4 mg), glutamic acid (156.5 mg), serine (50.4 mg) and aspartic acid (91.7 mg). Other amino acids present in less proportion are histidine (21.9 g), threonine (33.1 mg), tyrosine (30.6 mg), methionine (28.0 mg), hydroxyproline (26 mg), proline (0.5 mg) and sarcosine (0.70 mg). Minerals reportedly found in maca were iron (16.6 mg/100 g dry matter), calcium (150 mg/100 g dry matter), copper (5.9 mg/100 g dry matter), zinc (3.8 mg/100 g dry matter) and potassium (2050 mg/100 g dry matter) among others. [See 5].

Secondary Metabolites

Maca contains several secondary metabolites [5]. The secondary metabolites macaene and macamides are only found in this plant [16]. Macaenes are unsaturated fatty acids [16]. Other compounds include sterols as beta-sitosterol, campesterol and stigmasterol.

Different glucosinolates such as the aromatic glucosinolate glucotropaeolin has been identified in maca. Benzyl glucosinolate has been suggested as chemical marker for maca biological activity. However, this concept has been discarded since glucosinolates may easily metabolize to isothyocyanates and other smaller metabolites [17].

Benzyl glucosinolate is also present in another Peruvian plant named mashua (*Tropaelum tuberosum*). This plant however seem to possess effects opposite to that of maca, since its administration to male rats reduced sperm count [18] in contrast with the known effect of maca increasing sperm count [13].

It has been observed that maca batches from different producers vary significantly in the amount of macaene, macamides, sterols, glucosinolates [19-21]. In 2005, a publication appeared indicating that maca of different color types have different properties [12]. More recently, it has been found that maca color type associates with variations in concentrations of distinct bioactive metabolites [21, 22]. A difference in the proportion of secondary metabolites between maca color types may explain the variation in their biological properties.

Experimental Studies

Since 2000, several studies have been reported on biological effect of maca on experimental animals. These are summerized in Table 1.

It is important to note that differences in the method of preparation of maca extracts may yield biological effects. Traditionally maca is boiled or extracted in alcohol before being consumed [4]. In experimental studies, aqueous extract of maca is only effective after boiling pulverized maca hypocotyls in water.

The greatest effect on spermatogenesis was observed using the ethyl acetate fraction of the hydro-alcoholic extract of black maca [42]. Aqueous extract after boiling has a similar effect to that of hydro-alcoholic extract of maca [17]. Both aqueous and hydro-alcoholic extract of red maca reduced the prostate weight in rats with prostatic hyperplasia to a similar extent induced by testosterone enanthate (TE) [17]. There is considerable interest in the use of medicinal plants for the treatment of sexual disorders. Treatments of experimental animals with pulverized maca hypocotyls in doses of 15, 25, 75, and 100 mg/kg and the assessment of sexual behavior at 1, 7, 15, and 21 days of treatment yielded different results [23, 25]. The first study found increased sexual behavior of males at treatment days 1 and 15 [23], whereas the second study did not find changes in male sexual behavior at treatment days 1 or 21 [25]. Further studies are required to clarify these discrepancies. Testosterone controls sexual desire and spermatogenesis. However, the effect of maca on these physiological processes does not seem to be regulated by changes in serum testosterone or intra testicular testosterone levels. An alternative explanation would be that maca affects the androgen receptor . However, another study noted that maca did not modulate androgen receptors [43]. Thus, further chemical and molecular research is required to identify which of the many components of maca accounts for the effects observed.

Serum estradiol levels were also not affected in different studies which used mice [44], rats [13, 45], or humans [46]. Moreover, with an *in vitro* assay in our laboratory, we could not show that maca has a proliferative effect on MCF-7 cells [Vaisberg and Gonzales, unpublished observations].

Table 1. Properties for maca after in vivo administration in experimental animals

Species	Property	Source
Rats	Increase sperm count and sperm motility	[13]
	Increase male sexual behavior	[16, 23,24]
	small effect of rat male sexual behavior	[25]
	Nutritional effect	[26]
	Anti-stress properties	[27,28]
	Prevent testosterone-induced prostatic hyperplasia	[12]
	Reversed osteoporosis	[29,30]
	Neuroprotective effects	[31]
	Protect against UV radiation	[32]
	Improved antioxidant, lipid and glucose metabolism status	[33]
Mice	Increase male sexual behavior	[16]
	Increase embryo survival	[34]
	Prevent testosterone-induced prostatic hyperplasia	[35]
	Increased number of off springs	[34]
	Improves memory and learning	[14,36,37]
Guinea pigs	Increased number of off springs	[38]
Fish	Nutritional effect	[39,40]
	Increase embryo survival	[39]
Bulls	improve sperm quantity and quality Unaffected mating behavior	[41]

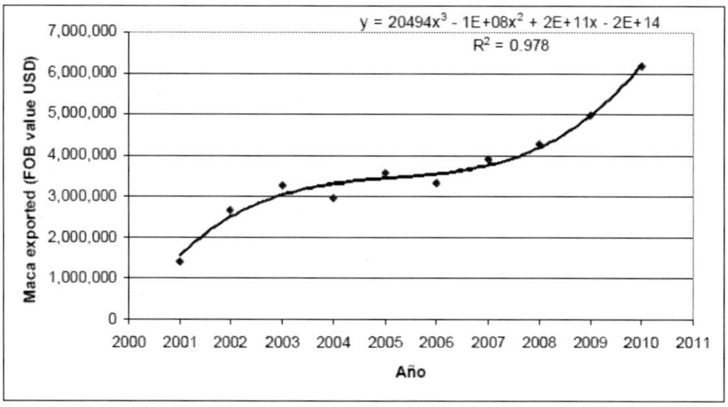

Figure 2. Maca exported from Peru in the last 10 years. Data are in FOB values (USD).

Studies in Humans

Interest for maca has been increased worldwide during the last 10 years. This is observed in Figure 2 which presents data on maca export from Peru, the only country producing maca. During 2010, Peru exported maca for a value of 6'179,011.8 USD, 4.36 times higher than value exported during 2001.

This increased demand for maca has also been accompanied by some concern about its safety. Piacente et al. [47] described the presence of (1R,3S)-1-methyl-1,2,3,4-tetrahydro-β-carboline-3-carboxylic acid (MTCA) in maca hypocotyls.

On this finding, the authors made some generalizations about the action of MTCA suggesting that it can be toxic [47]. These affirmations have motivated the French Agency for Sanitary Security (AFFSA) to issue an opinion about the health risk for the consumers of pulverized roots of maca [48]. However, MTCA also occurs on fruits like oranges and grapefruit and fruit juices [49], which are frequently used because of their favorable properties on health. MTCA has been described on the fermented garlic extract [50,51] and its concentration increases with time and in turn increasing its antioxidant activity.

Moreover, MTCA is also detected in several foods and in some, its concentrations relatively high (greater than the ones found by Piacente in maca) suggesting that claims are overestimated.

In a recent paper, several arguments indicate that MTCA in maca is safe [52].

In addition, maca is not mutagenic and it also contains several beneficial compounds, some of which has anti carcinogenic properties [5, 53]. We feel that consumption of maca must not generate any concern as mentioned in the French alert [48], since no toxicity has been reported with traditional maca consumption that requires a boiling process. MTCA is a natural constituent of several plants and upon consumption of such plants there is no toxicity was reported. This suggests that maca as a whole multi-component source may not demonstrate any adversity action upon consumption.

Furthermore, a recent study was designed to investigate health status in a population from the Peruvian central Andes (Carhuamayo, 4100 m) which traditionally is consuming maca and compared with a population from the same area which does not consume maca. The study based on a survey, assessed maca consumption, socio-demographic aspects, health status, and fractures in men and women aged 35-75 years old. In a sub-sample their hepatic and kidney functions and hemoglobin values were assessed. From the sample studied, 80% of the population consumed maca. The 85% of them consume maca for nutritional purposes. These populations have used maca since childhood mainly after hypocotyls is naturally dried. The consumption is mainly as juices, and the variety that they consume are a mixture of different colors of the hypocotyls. Maca consumption was associated with higher score in health status, lower rate of fractures, lower scores of sign and symptoms of chronic mountain sickness.

In addition, maca consumption was associated with low body mass index and low systolic blood pressure. Hepatic and kidney function, lipid profile and glycemia were normal in the population consuming maca. In conclusion, this study demonstrated in a population traditionally using maca the consumption of maca is safe [4].

Maca and Sexual Function

Sexual dysfunctions are highly prevalent in our society worldwide, and that the occurrence of sexual dysfunctions increases directly with age for both men and women [54]. They occur in 20%-30% of men and 40-45% of women according to 18 descriptive epidemiological studies from around the world [55]. Most sexual problems relate to sexual desire (interest in sex) in both females and males and male erectile dysfunction (ED) [55]. Interest for medicinal plants to treat sexual dysfunctions has been increased in the last 20 years [56].

Maca has been described to improve sexual behavior in experimental animals [16,23,and 24] although conflictive results may be observed [25]. Traditionally maca has been referred as a plant to improve fertility [2] and as energizer [3]. In a randomized study we were unable to demonstrate effect of maca on penile erection in apparently healthy adult men after 12 weeks of treatment with gelatinized maca compared with results using placebo [Gonzales, unpublished data].

Recently, a systematic review has been performed on effect of maca on sexual function in humans [57]. In this review, only four randomized clinical trials (RCT) met all the inclusion criteria [58-61]. According the review, two RCTs suggested a significant positive effect of maca on sexual dysfunction or sexual desire in healthy menopausal women [58] or healthy adult men [59], respectively, while the other RCT according to the reviewers failed to show any effects in healthy cyclists.

However, analyzing results from such study, authors showed that maca extract significantly improved the self-rated sexual desire score compared to the baseline test (P=0.01), and compared to the placebo trial after supplementation (P=0.03) [60]. In this study the effect was evident as early as 14 days of treatment which is significantly shorter that that showed with gelatinized maca in which effects were observed only after 8 weeks of treatment.

The RCT further assessed the effects of maca in patients with mild erectile dysfunction using the International Index of Erectile Dysfunction-5 and showed significant effects on subjective perception of general and sexual well-being.[61]. One study was not included in the systematic review because no placebo effect was assessed in their research protocol [62].

In that study, maca was administered in two doses (1.5 g/day and 3-0 g/day) to patients with selective-serotonin reuptake inhibitor (SSRI)-induced sexual dysfunction. The Arizona Sexual Experience Scale (ASEX) and the Massachusetts General Hospital Sexual Function Questionnaire (MGH-SFQ) were used to measure sexual dysfunction. Subjects on 3.0 g/day maca had a significant improvement in ASEX (from 22.8 ± 3.8 to 16.9 ± 6.2; z=-2.20, P=0.028) and in MGH-SFQ scores (from 24.1 ± 1.9 to 17.0 ± 5.7; z=-2.39, P=0.017), but subjects on 1.5 g/day maca did not. Libido improved significantly (P<0.05) based on ASEX item #1, but not by dosing groups. Maca was well tolerated by the subjects in that study[62].

Although evidence suggests an effect of maca on sexual desire and mild erectile dysfunction data also revealed that maca extract seems to have better effect [60] than gelatinized maca [59] or maca powder.

Maca and Sperm Function

In a study of 9 apparently healthy men who had received maca for 4 months showed an increase in seminal volume, sperm count and sperm motility [63] (Table 2). Serum hormone levels (LH, FSH, prolactin, estradiol and testosterone) in men were not affected by treatment with maca [63,64].

Maca powder and maca extract were unable to activate androgen receptor-mediated transcription in prostate cancer cell lines [43] or in a yeast-based hormone-dependent reporter assay [58].

Maca As Energizer

Maca has been shown to reduce in depression and anxiety based on inventory scores [58, 62]. A self perception survey showed that maca acted as energizer compared with placebo in men apparently healthy [3]. Maca extract administration for 14 days significantly improved 40 km cycling time performance compared to the baseline test (P=0.01), but not compared to the placebo trial after supplementation (P>0.05).

Table 2. Semen variables before and 4-month after maca treatment

Semen variable	Pre-maca N=9	Post-maca N=9	*P* value
Volume (mL)	2.23±0.28	2.91±0.28	<0.05
pH	7.47±0.09	7.44±0.07	NS
Sperm count (10^6/mL)	67.06±18.61	90.33±20.46	NS
Total sperm count (10^6/mL)	140.95±31.05	259.29±68.17	<0.05
Motile sperm count (10^6/mL)	87.72±19.87	183.16±47.84	<0.05
Sperm motility grade a (%)	29.00±5.44	33.65±3.05	NS
Sperm motility grade a+b (%)	62.11±3.64	71.02±2.86	<0.05
Normal sperm morphology (%)	75.50±2.02	76.90±1.23	NS

Source: [63]. Data are mean ± standard error of the mean. N= number of subjects: NS: Not significant.

Maca and Metabolic Syndrome

One study reported on effects of maca alone or combined with another supplements in patients with metabolic syndrome. The randomized placebo-controlled 90-day study assessed the effects of maca and yacon in combination with silymarin on plasma and lipoprotein lipids, serum glucose and safety parameters in patients suffering from the metabolic syndrome. No adverse effects were found in volunteers using silymarin (0.8 g/day), silymarin+yacon (0.8+2.4 g/day) and silymarin+maca (0.6+0.2 g/day). A moderate AST level and diastolic blood pressure increase was found in volunteers using maca (0.6 g/day) [65].

However, in a randomized clinical trial in healthy men showed that gelatinized maca reduced systolic and diastolic blood pressure after 12 weeks of treatment [3]. Moreover, Maca inhibited significantly the hypertension relevant angiotensin I-converting enzyme (ACE) *in vitro* [66]. In a population traditionally consuming maca, systolic blood pressure was lower than in those not consuming maca [4]. However, AST levels were similar in those consuming and those no consuming maca [4]. Maca contains high amounts of potassium [5]. Potassium is an important nutrient to reduce risk of hypertension [67]

Maca and Osteoarthritis

In a randomized double-blind study on 95 patients with osteoarthritis, a combination of *Uncaria gianensis (*cat´s claw*)* (300 mg) and maca (1,500 mg) was administered twice daily for 8 weeks and compared to a treatment with glucosamine sulfate. Both treatments substantially improved pain, stiffness, and functioning in the patients [68]. However, as the study did not include a placebo control group, glucosamine effects remain unclear.

Toxicity

Maca has been used for centuries in the Central Andes of Peru and, no toxic effects have been reported if it was consumed after boiling [5]. Previous review data on *in vivo* and *in vitro* studies with maca indicate that its use is safe [5]. Further evidence shows that aqueous and methanolic extracts of maca do not display *in vitro* hepatotoxicity [69]. Moreover, freeze dried aqueous extract of maca (1 g/kg BW) in mice did not reveal any toxic effect on the normal development of pre-implanted mouse embryos [70].

Results in rats show that different types of maca (black, red, yellow) have no acute toxicity at ≤17g of dried hypocotyls/kg BW. Rats treated chronically for 84 days with 1 g/Kg BW showed no side effects and their liver histology was similar to that observed in controls [71]. Since the standard doses in rats for maca are in the range of 1–2 g/Kg BW, this further suggests that maca is safe. Human consumption of ≤1 g/kg per day is considered safe, as well. However, as discussed above in a study in patients with metabolic syndrome the administration of maca at a dose of 0.6 g/day for 90 days resulted in a moderate elevation of AST and diastolic arterial pressure [65]. This has not been confirmed in other studies [3,4]. Data on population of 600 subjects in the Peruvian central Andes showed that maca consumption was safe and that health status was improved [4].

Conclusion

Consumption of Maca has significantly increased during the last 10 years worldwide (Figure 2). Clearly, further research is required to address the mechanisms of actions and the active principles of this plant. However, available data suggest that maca has several important biological properties, and scientific evidence of these properties could be important for farmers, dealers, and consumers. Furthermore, it is necessary to demonstrate the biological effects of specific secondary metabolites of maca and their actions when added as a mixture.

Maca is a plant with great potential as an adaptogen and appears to be very promising as a source for nutraceuticals in the prevention of several diseases. Scientific evidence has shown the effects of maca on sexual behavior, fertility, mood, memory, osteoporosis, metabolism and the treatment of some tumor entities. However, the active principles responsible for each of these effects are still unknown. Macamides have been described as novel compounds of maca that have not been found in any other plant species so far [16]. It is suggested that this lipid fraction of maca may be responsible for the increase in sexual behavior [16, 24]. Studies on testicular function, spermatogenesis, fertility, mood, memory, and prostatic hyperplasia [13, 14, 72, and 73] were performed with aqueous extracts that contain only trace amounts of macamides [19]. This suggests that compounds other than macamides are responsible for these activities. It is important to note that no studies of nutrigenomic effects have been done on maca and its chemical constituents providing an excellent opportunity to explore maca as a nutraceutical source.

References

[1] Quiroz, C., Aliaga, R.: Maca (*Lepidium meyenii* Walp.); In: Hermann, M., Hellers, J. (eds): Andean Roots and Tubers: Ahipa, Arracacha, Maca and Yacon. Promoting the Conservation and Use of Underutilized Neglected Crops. 21. Rome, *International Plant Genetic Resources Institute,* 1997, pp 173–197.

[2] Cobo, B. [1653]: *History of the New World*. Madrid, Biblioteca de Autores Españoles, 1956.

[3] Gonzales, G. F.: Biological effects of *Lepidium meyenii*, maca, a plant from the highlands of Peru; In: Singh, V. K., Bhardwaj, R., Govil, J. N., Sharma, R. K. (Eds): *Natural Products. Series: Recent Progress in Medicinal Plants*, Vol. 15. Houston, TX, Studium Press, 2006, pp 209–234.

[4] Gonzales, G. F. MACA: Del alimento perdido de los Incas al milagro de los Andes: Estudio de seguridad alimentaria y nutricional. *Segurança Alimentar e Nutricional, Campinas*, 16 17 (1): 16-36, 2010.

[5] Valerio, L., Gonzales, G. F.: Toxicological aspects of South American herbs: *Uncaria tomentosa* (Cat's Claw) and *Lepidium meyenii* (maca). A critical synopsis. *Toxicol. Rev*. 2005;24:11–35.

[6] Cieza de León, P.: *Chronicle of Peru*. First Part. London, Hakluyt Society, 1553.

[7] Ruiz, H.: Relación histórica del viaje a los reinos del Perú y Chile, 1777–1778. Madrid, Academia de Ciencias Exactas, *Físicas y Naturales*, 1952.

[8] Oerlemans, K., Barrett, D. M., Suades, C. B., Verkerk, R., Dekker, M.: Thermal degradation of glucosinolates in red cabbage. *Food Chemistry*. 2006; 95:19–29.
[9] Matusheski, N. V., Juvik, J. A., Jeffery, E. H.: Heating decreases epithiospecifier protein activity and increases sulforaphane formation in broccoli. *Phytochemistry*. 2004;65:1273–1281
[10] Dewanto, V., Wu, X., Adom, K. K., Liu, R. H.: Thermal processing enhances the nutritional value of tomatoes by increasing total antioxidant activity. *J. Agric .Food Chem.* 2002;50: 3010–3014.
[11] Tello, J., Hermann, M., Calderón, A.: La maca (*Lepidium meyenii* Walp.) cultivo alimenticio potencial para las zonas altoandinas. *Bol. Lima.* 1992;14:59–66.
[12] Gonzales, G. F., Miranda, S., Nieto, J., Fernandez, G., Yucra, S., Rubio, S., Yi, P., Gasco, M.: Red maca (*Lepidium meyenii*) reduced prostate size in rats. *Reprod. Biol. Endocrinol.* 2005;3:5.
[13] Gonzales, C., Rubio, J., Gasco, M., Nieto, J., Yucra, S., Gonzales, G. F.: Effect of short term and long term treatments with three ecotypes of *Lepidium meyenii* (maca) on spermatogenesis in rats. *J. Ethnopharmacol.* 2006;103:448–454.
[14] Rubio, J., Caldas, M., Dávila, S., Gasco, M., Gonzales, G. F.: Effect of three different cultivars of *Lepidium meyenii* (maca) on learning and depression in ovariectomized mice. *BMC Complement Altern. Med.* 2006; 6:23.
[15] Dini, A., Migliuolo, G., Ratrelli, L., Saturnino, P., Schettino, O.: Chemical composition of *Lepidium meyenii. Food Chem.* 1994; 49:347–349.
[16] Zheng, B. L., He, K., Kim, C. H., Rogers, L., Yu, S., Huang, Z. Y., Lu, Y., Yan, S. J., Qien, L. C., Zhen, Q. Y.: Effect of a lipidic extract from *Lepidium meyenii* on sexual behavior in mice and rats. *Urology*. 2000; 55:598–602.
[17] Gonzales, G. F., Vásquez, V., Rodríguez, D., Maldonado, C., Mormontoy, J., Portella, J., Pajuelo, M., Villegas, L., Gasco, M.: Effect of two different extracts of red maca in male rats with testosterone-induced prostatic hyperplasia. *Asian J. Androl.* 2007;9: 245–251.
[18] Cárdenas-Valencia, I., Nieto, J., Gasco, M., Gonzales, C., Rubio, J., Portella, J., Gonzales, G. F. (2008) Tropaeolum tuberosum (Mashua) reduces testicular function: effect of different treatment times. *Andrologia*. 40:352–357.
[19] McCollom, M. M., Villinski, J. R., McPhail, K. L., Craker, L. E., Gafner, S.: Analysis of macamides in samples of Maca (*Lepidium meyenii*) by HPLC-UV-MS/MS. *Phytochem. Anal.* 2005;16: 463–469.
[20] Ganzera, M., Zhao, J., Muhammad, I., Khan, I. A. Chemical profiling and standardization of Lepidium meyenii (maca) by reversed phase high performance liquid chromatography. *Chem. Pharm. Bull.* (Tokyo). 2002; 50: 988-991.
[21] Clément, C., Díaz, D., Manrique, I., Avula, B., Khan, I. A., Ponce-Aguirre, D., Kunz, C., Mayer, A., Kreuzer, M. Secondary metabolites in maca as affected by hypocotyl color, cultivation history, and site. *Agronomy J.* 2010; 102: 431-439.
[22] Clément, C., Díaz, D., Avula, B., Khan, I. A., Ponce-Aguirre, D., Manrique, I., Kreuzer, M. Influence of colour type and previous cultivation on secondary metabolites in hypocotyls and leaves of maca (*Lepidium meyenii* Walpers). *J. Sci. Food Agric.* 2010; 90: 861-869.

[23] Cicero, A. F., Bandieri, E., Arletti, R.: *Lepidium meyenii* Walp. improves sexual behaviour in male rats independently from its action on spontaneous locomotor activity. *J. Ethnopharmacol.* 2001;75: 225–229.

[24] Cicero, A. F., Piacente, S., Plaza, A., Sala, E., Arletti, R., Pizza, C.: Hexanic maca extract improves rat sexual performance more effectively than methanolic and chloroformic maca extracts. *Andrologia.* 2002;34: 177–179.

[25] Lentz, A., Gravitt, K., Carson, C. C., Marson, L. Acute and chronic dosing of Lepidium meyenii (Maca) on male rat sexual behavior. *J. Sex. Med.* 2007;4:332-339

[26] Canales, M., Aguilar, J., Prada, A., Marcelo, A., Huaman, C., Carvajal, L.: Nutritional evaluation of *Lepidium meyenii* (maca) in albino mice and their descendants [in Spanish]. *Arch. Latinoam. Nutr.* 2000; 50: 126–133.

[27] Lopez-Fando, A., Gomez-Serranillos, M. P., Iglesias, I., Lock, O., Upamayta, U. P., Carretero, M. E.: *Lepidium peruvianum* chacon restores homeostasis impaired by restraint stress. *Phytother. Res.* 2004;18: 471–474.

[28] Tapia, A., López, C., Marcelo, A., Aguilar, J. L.: The Maca (*Lepidium meyenii*) and their effect anti-stress in an animal model in mice [in Spanish]. *Acta Andina.* 2000;8:45–56.

[29] Gonzales, C., Cárdenas-Valencia, I., Leiva-Revilla, J., Anza-Ramirez, C., Rubio, J., Gonzales, G. F. Effect of different varieties of Maca (*Lepidium meyenii*) on bone structure in ovariectomized rats. *Research in Complementary Medicine.* 2010; 17: 137-143.

[30] Zhang, Y., Yu, L., Ao, M., Jin, W. Effect of ethanol extract of Lepidium meyenii Walp. on osteoporosis in ovariectomized rat. *J. Ethnopharmacol.* 2006; 105274-279.

[31] Pino-Figueroa, A., Nguyen, D., Maher, T. J. Neuroprotective effects of Lepidium meyenii (Maca). *Ann. N. Y. Acad. Sci.* 2010; 1199:77-85.

[32] Gonzales-Castañeda, C., Gonzales, G. F. Hypocotyls of *Lepidium meyenii* (maca), a plant of the Peruvian highlands, prevents the ultraviolet A, B and C (UVA, UV B and UV C)-induced skin damage in rats. *Photodermatology, Photoimmunology and Photomedicine.* 2008; 24: 24-31.

[33] Vecera, R., Orolin, J., Skottová, N., Kazdová, L., Oliyarnik, O., Ulrichová, J., Simánek, V. The influence of maca (Lepidium meyenii) on antioxidant status, lipid and glucose metabolism in rat. *Plant Foods Hum. Nutr.* 2007 Jun.; 62(2):59-63.

[34] Ruiz-Luna, A. C., Salazar, S., Aspajo, N. J., Rubio, J., Gasco, M., Gonzales, G. F. *Lepidium meyenii* (Maca) increases litter size in normal adult female mice. *Reproductive Biology and Endocrinology.* 2005; 3(1):16

[35] Gonzáles, G. F., Gasco, M., Malheiros-Pereira, A., Gonzales-Castañeda, C. Antagonistic effect of *lepidium meyenii* (red maca) on prostatic hyperplasia in adult mice. *Andrologia.* 2008;40: 179-185.

[36] Rubio, J., Dang, H., Gong, M., Liu, X., Chen, S.-L., Gonzales, G. F. Aqueous and hydroalcoholic extracts of Black Maca (*Lepidium meyenii*) improve scopolamine-induced memory impairment in mice. *Food and Chemical Toxicology.* 2007; 45:1882-1890.

[37] Rubio, J., Qiong, W., Liu, X., Jiang, Z., Dang, H., Chen, S. L., Gonzales, G. F. Aqueous Extract of Black Maca (Lepidium meyenii) on Memory Impairment Induced by Ovariectomy in Mice. *Evid. Based Complement Alternat. Med.* 2011 (2011), Article ID 253958, 7 pages

[38] Alvarez, C. J.: Utilización de diferentes niveles de Maca en la fertilidad de cobayos. Bachellor Thesis. Faculty of Agriculture and Cattle Sciences, Universidad Nacional Daniel Alcides Carrión, Pasco, 1993.

[39] Lee, K.-J., Dabrowski, K., Rinchard, J., Gomez, C., Guz, L., Vilchez, C.: Supplementation of maca (*Lepidium meyenii*) tuber meal in diets improves growth rate and survival of rainbow trout *Oncorhynchus mykiss* (Walbaum) alevins and juveniles. *Aquaculture Res.* 2004; 35:215–223.

[40] Lee, K.-J., Dabrowski, K., Sandoval, M., Miller, M. J. S.: Activity-guided fractionation of phytochemicals of maca meal, their antioxidant activities and effects on growth, feed utilization, and survival in rainbow trout (*Oncorhynchus mykiss*) juveniles. *Aquaculture.* 2005;244:293–301.

[41] Clément, C., Kneubühler, J., Urwyler, A., Witschi, U., Kreuzer, M. Effect of maca supplementation on bovine sperm quantity and quality followed over two spermatogenic cycles. *Theriogenology.* 2010; 74:173-83.

[42] Yucra, S., Gasco, M., Rubio, J., Nieto, J., Gonzales, G. F.: Effect of different fractions from hydroalcoholic extract of black maca (*Lepidium meyenii*) on testicular function in adult male rats. *Fertil. Steril.* 2008; 89(5 suppl):1461–1467.

[43] Bogani, P., Simonini, F., Iriti, M., Rossoni, M., Faoro, F., Poletti, A., Visioli, F.: *Lepidium meyenii* (maca) does not exert direct androgenic activities. *J. Ethnopharmacol.* 2006; 104:415–417.

[44] Oshima, M., Gu, Y., Tsukada, S.: Effects of *Lepidium meyenii* Walp and *Jatropha macrantha* on blood levels of estradiol-17β, progesterone, testosterone and the rate of embryo implantation in mice. *J. Vet. Med. Sci.* 2003; 65:1145–1146.

[45] Gasco, M., Villegas, L., Vilegas, L., Rubio, J., Gonzales, G. F.: Dose-response effect of red maca (*Lepidium meyenii*) on benign prostatic hyperplasia induced by testosterone enanthate. *Phytomedicine.* 2007; 14:460–464.

[46] Gonzales, G. F., Córdova, A., Vega, K., Chung, A., Villena, A.: Effect of *Lepidium meyenii* (maca), a root with aphrodisiac and fertility-enhancing properties, on serum reproductive hormone levels in adult healthy men. *J. Endocrinol.* 2003; 176:163–168.

[47] Piacente, S., Carbone, V., Plaza, A., Zampelli, A., Pizza, C. Investigation of the tuber constituents of maca (Lepidium meyenii Walp). *J. Agric. Food Chem.* 2002; 50: 5621-5625.

[48] APSSA. From the French Agency of Sanitary Security on foods relative to the risk assessment for health by consuming pulverized maca roots or as alimentary supplement. *Afssa-Saisine.* 2004-SA-0155: 1-3.

[49] Herraiz, T., Galisteo, J. Tetrahydro-beta-carboline alkaloids occur in fruits and fruit juices. Activity as antioxidants and radical scavengers. *J. Agric. Food. Chem.* 2003; 51: 7156-7161.

[50] Sato, E., Kohno, M., Niwano, Y. Increased level of tetrahydro-beta-carboline derivatives in short-term fermented garlic. *Plant Foods Hum. Nutr.* 2006; 61:175-178.

[51] Ichikawa, M., Yoshida, J., Ide, N., Sasaoka, T., Yamaguchi, H., Ono, K. Tetrahydro-beta-carboline derivatives in aged garlic extract show antioxidant properties. *J. Nutr.* 2006; 136 (3 Suppl.):726S-731S.

[52] Gonzales, G. F., Gonzales-Castañeda, C. The Methyltetrahydro-*b*-Carbolines in Maca (*Lepidium meyenii*). *Evid. Complem. Altern. Med.* 2009; 6:315-6.

[53] Gonzales, G. F., Valerio, L. G. Jr. Medicinal Plants from Peru: A review of plants as potential agents against cancer. *Anti-Cancer Agents Med. Chem.* 2006; 6: 429-444.

[54] Derogatis, L. R., Burnett, A. L. The epidemiology of sexual dysfunctions. *J. Sex. Med.* 2008;5:289-300.

[55] Lewis, R. W., Fugl-Meyer, K. S., Corona, G., Hayes, R. D., Laumann, E. O., Moreira, E. D. Jr, Rellini, A. H., Segraves, T. Definitions/epidemiology/risk factors for sexual dysfunction. *J. Sex. Med.* 2010;7: 1598-607.

[56] McKay, D. Nutrients and botanicals for erectile dysfunction: examining the evidence. *Altern. Med. Rev*. 2004; 9:4-16.

[57] Shin, B. C., Lee, M. S., Yang, E. J., Lim, H. S., Ernst, E. Maca (L. meyenii) for improving sexual function: a systematic review. *BMC Complement Altern. Med.* 2010; 10:44.

[58] Brooks, N. A., Wilcox, G., Walker, K. Z., Ashton, J. F., Cox, M. B., Stojanovska, L: Beneficial effects of *Lepidium meyenii* (Maca) on psychological symptoms and measures of sexual dysfunction in postmenopausal women are not related to estrogen or androgen content. *Menopause*. 2008; 15:1157–1162.

[59] Gonzales, G. F., Cordova, A., Vega, K., Chung, A., Villena, A., Gonez, C., Castillo, S.: Effect of *Lepidium meyenii* (maca) on sexual desire and its absent relationship with serum testosterone levels in adult healthy men. *Andrologia*. 2002; 34:367–372.

[60] Stone, M., Ibarra, A., Roller, M., Zangara, A., Stevenson, E. A pilot investigation into the effect of maca supplementation on physical activity and sexual desire in sportsmen. *J. Ethnopharmacol*. 2009 Dec. 10; 126:574-576.

[61] Zenico, T., Cicero, A. F., Valmorri, L., Mercuriali, M., Bercovich, E.: Subjective effects of *Lepidium meyenii* (maca) extract on well-being and sexual performances in patients with mild erectile dysfunction: a randomized, double-blind clinical trial. *Andrologia*. 2009;41:95–99.

[62] Dording, C. M., Fisher, L., Papakostas, G., Farabaugh, A., Sonawalla, S., Fava, M., Mischoulon, D.: A doubleblind, randomized, pilot dose-finding study of maca root (*L. meyenii*) for the management of SSRI induced sexual dysfunction. *CNS Neurosci. Ther.* 2008; 14:182–191.

[63] Gonzales, G. F., Córdova, A., Gonzales, C., Chung, A., Vega, K., Villena, A.: Improved sperm count after administration of *Lepidium meyenii* (maca) in adult men. *Asian J. Androl*. 2001; 3:301–304.

[64] Gonzales, G. F., Córdova, A., Vega, K., Chung, A., Villena, A.: Effect of *Lepidium meyenii* (maca), a root with aphrodisiac and fertility-enhancing properties, on serum reproductive hormone levels in adult healthy men. *J. Endocrinol*. 2003;176:163–168.

[65] Valentová, K., Stejskal, D., Bartek, J., Dvorácková, S., Kren, V., Ulrichová, J., Simánek, V. Maca (Lepidium meyenii) and yacon (Smallanthus sonchifolius) in combination with silymarin as food supplements: in vivo safety assessment. *Food Chem. Toxicol*. 2008; 46:1006-1013.

[66] Ranilla, L. G., Kwon, Y. I., Apostolidis, E., Shetty, K. Phenolic compounds, antioxidant activity and in vitro inhibitory potential against key enzymes relevant for hyperglycemia and hypertension of commonly used medicinal plants, herbs and spices in Latin America. *Bioresour. Technol*. 2010; 101:4676-4689.

[67] Lee, J. S., Park, J., Kim, J. Dietary factors related to hypertension risk in Korean adults-data from the Korean national health and nutrition examination survey III. *Nutr. Res. Pract.* 2011; 5:60-65.

[68] Mehta, K., Gala, J., Bhasale, S., Naik, S., Modak, M., Thakur, H., Deo, N., Miller, M. J.: Comparison of glucosamine sulfate and a polyherbal supplement for the relief of osteoarthritis of the knee: a randomized controlled trial [ISRCTN25438351]. *BMC Complement Altern. Med.* 2007; 7:34.

[69] Valentova, K., Buckiova, D., Kren, V., Peknicova, J., Ulrichova, J., Simanek, V.: The in vitro biological activity of *Lepidium meyenii* extracts. *Cell Biol. Toxicol.* 2006;22:91–99.

[70] D'Arrigo, G., Benavides, V., Pino, J.: Preliminary evaluation effect of *Lepidium meyenii* Walp on the embryo development of mouse. *Rev. Per. Biol.* 2004;11:103–106.

[71] Gasco, M., Aguilar, J., Gonzales, G. F. Effect of chronic treatment with three varieties of Lepidium meyenii (Maca) on reproductive parameters and DNA quantification in adult male rats. *Andrologia.* 2007; 39:151-158.

[72] Chung, F., Rubio, J., Gonzales, C., Gasco, M., Gonzales, G. F.: Dose-response effects of *Lepidium meyenii* (maca) aqueous extract on testicular function and weight of different organs in adult rats. *J. Ethnopharmacol.* 2005; 98:143–147.

[73] Gonzales, G. F., Gasco, M., Córdova, A., Chung, A., Rubio, J., Villegas, L.: Effect of *Lepidium meyenii* (maca) on spermatogenesis in male rats acutely exposed to high altitude (4340 m). *J. Endocrinol.* 2004; 180:87–95.

Bioethical and Educational Issues in Nutrigenomics

In: Nutrigenomics
Editor: M. T. Ravi Subbiah
ISBN: 978-1-62417-051-5

Chapter XII

Application of Nutrigenomics in Personalized Nutrition: Other Critical Issues

M. T. Ravi Subbiah
Department of Internal Medicine,
University of Cincinnati Medical Center, Cincinnati, Ohio, US

Abstract

Nutrigenomics has the potential to deliver tools for personal management of health issues. Also, there is considerable public demand to know personal genetic information (PGI) regarding their susceptibility for chronic diseases. This is bringing in a fundamental shift in how the public is using this information to change their diet and life style. Furthermore, availability of high throughput technologies has propelled the formation of companies which offer genetic services to individuals directly on-line and advocate dietary advice based on some tests. This raises a number of ethical issues in terms of consumer protection and privacy. Some of the questions raised include: do these tests have clinical validity? Are the tests done using quality control conditions that are comparable to clinical tests done in a medical facility? Is it legal for a company that offers tests, also to advocate dietary advice? Personalized nutrition may also change the way food is viewed, based on its ability to prevent diseases. Lastly, it is critical to develop more professionals in this rapidly advancing field.

Keywords: **n**utrigenomics, personalized medicine, bioethical issues, nutraceuticals, functional foods, dermagenetics, cosmeceuticals, direct-to-consumer genetics

Introduction

In recent years, there is considerable public demand to know personal genetic information(PGI) regarding their susceptibility for chronic diseases and other phenotypic traits. This is bringing in a fundamental shift in how the public is using this information to change their diet and life style [1]. Furthermore, availability of high throughput technologies has propelled the formation of companies which offer genetic services to individuals directly on-line and advocate dietary advice based on some tests. This raises a number of ethical issues in terms of consumer protection and privacy. Three major issues that need attention for the success of nutrigenomics in personalized medicine are a) direct-to-consumer genetics (DTCG); b) bioethical and regulatory aspects related to individualized food habits, nutraceuticals and functional foods and c) nutrigenomics education and training for professionals.

Direct-to-consumer Genetics (DTCG)

There are many companies in the USA through which the public can get PGI by direct on-line ordering of genetic tests (with a sample of cheek swabs). Some examples of genetic testing companies are Genewise and Genelink (Longwood, FL), 23 and Me (Mountain View, CA) and Navigenics (Foster City, CA). There is considerable debate whether the consumers are really getting valid information that can predict health status and disease risk from these companies. Saukko et. al [2] has carefully delineated the value of nutrigenomic tests that are offered, their interpretation and place in medicine/consumer culture/society boundaries. One of the concerns is that there may be considerable disparity between what the consumer imagines the test will offer and its actual value in predicting disease risk. Recently, Cervinski and Tsongalis [3] have collected critical insights into the advantages and concerns about DTCG based on their interviews with five experts. Some of the major advantages and concerns noted about DTCG are listed below:

Advantages:

1) The convenience and ready access of the on-line tests.
2) The personal and individualized attention the user receives.
3) Personal empowerment, a sense of more control and gaining information about themselves.
4) Eliminates the need to physically be present in a clinical facility in order to get the test done.
5) Public sees more options for preventive care in terms of using dietary supplements, nutraceuticals and functional foods.
6) Some tests related to the risk of chronic diseases which can indeed offer early diagnosis, can play a role in decreasing risk of debilitating disease and controlling long term escalating health care costs

Concerns:

1. One of the main concerns regarding DTCG is that both the sample collection and the complex test performance are done outside the central laboratory and health care facility (unlike all complex clinical testing done today).
2. It is not established that all tests have real clinical validity and utility.

3. It is doubtful that useful interpretation and personal guidance are provided following genetic analysis.
4. Removing the role of health care professionals in the process of on-line ordering of tests is a big concern.
5. It is not clear that the on-line laboratories follow adequate standards, appropriate quality controls and sampling issues so critical and rigidly enforced in clinical laboratory analysis.
6. Currently regulations in-place to enforce guidelines of quality control (exm: CLIA certification)for the on-line laboratories are not adequate.
7. It is not established whether some of the commonly used on-line tests (so called health/life style tests) actually matches with the claimed health risks . This is because, post-translational , epigenomic and environmental influences can alter disease risk resulting from a given genetic mutation. Therefore, it may be more useful to match-up a genetic mutation test with an actual change in a plasma metabolite. In other words, use of metabolomics as a valid tool to complement genetic testing may furnish more valid clinical information and meaningful dietary guidance.
8. There is a strong concern about privacy and safety of the genetic information such as, who will have access to the information and what happens to the sample and the data in the long run. Also, there are concerns whether guidelines for the Health Insurance Portability and Accountability Act (HIPAA) are followed strictly.

Bio-Ethical Issues

Nutrigenomics has the potential to deliver tools for personal management of health issues. This puts food choices and dietary habits in a totally new light. As indicated by Komduur et.al. [4], the food choices will depend upon its potential to decrease risk factors through food/gene interactions. Furthermore, food will be strictly viewed from its ability to prevent diseases. This may place an undue burden of affordability within some ethnic and economic groups raising ethical questions. Some of the bioethical implications of this concept and its validation by research have also been examined by Bergmann et.al [5]. These authors noted that the standards for ethical issues commonly encountered in any human research will be further challenged in genetic testing and research dealing with nutrigenomics. Furthermore, their analysis raises concern whether the practice of nutrigenomics is ready for prime time. Some of these concerns include: a) the benefit of nutrigenomics in bringing about improvement in health care and /or disease prevention is not clearly established yet; b) the validity of personal nutrition, a key concept driving nutrigenomics in terms of a genotype specific diet is premature and needs more research and c) the promotion of personalized nutrition may have some unwanted consequences such as medicalization of food and eating behavior.

Some of the concerns of individualization of health interventions as a consequence of nutrigenomic applications have also been discussed by Ozdemir and Godard [4]. These authors list several specific ethical led social issues (ELSI's) that needs attention. These include a) issues related to the privacy of genetic information and its implications for health insurance and employment and b) prevention/ modification focus of pre-disposed phenotypes in apparently healthy individuals, an approach that crosses conventional research boundaries. These issues can create new social norms, values, and responsibilities, newer life style products and the way we perceive disease itself [2, 5].

The above mentioned factors related to nutrigenomics and personalized nutrition practice and business will place heavy burden in defining and validating the claims of functional foods. Functional foods are defined as foods offering benefit to human health beyond the effects of nutrients alone and the hope is to match foods to the needs of individual genetype [6, 7]. One example of a functional food is Benecol (Raisio Nutrition Ltd, Finland), a margarine containing phytostanol esters [8]. This product was based on the numerous studies over the past 40 years which have shown that plant sterols inhibit cholesterol absorption [9] thereby lowering plasma cholesterol and help decrease cardiovascular risk. Further studies showed that phytostanol esters (saturated derivatives of phytosterols) are more effective in lowering cholesterol than unsaturated phytosterols. This product might be useful to individuals with a genetic mutation (genotype C1950G) in ABCG5, one of the ATP Binding Cassete Transporters, who demonstrate increased absorption of dietary cholesterol [10, 11].

There are three important issues concerning regulation of functional foods [12]. These are: 1) Safety, any concerns that systemically levels of food components might reach high levels, might be injurious to health or negate benefits of other food components; 2) Efficacy, should be able to document their functionality and effect and 3) Their long term effect, on over-all health. Guidelines for achieving uniformity in the validation of the claims associated with functional foods and to provide international criteria for clinical trials for assessment of evidence of claim for a food, are constantly being updated. Some of these criteria has been summarized by Aggett [13] and are listed below:

1. Establish chemical characterization of the food component responsible for the given metabolic effect.
2. Establish a dose/ response relationship
3. Account for confounders in the study
4. Study should be of appropriate duration to reach meaningful conclusions
5. Deal with compliance issues
6. Satisfactory statistical analysis (power etc.) to defend conclusions
7. Use of validated biomarkers for given outcome

All these criteria support an expectation that functional food should have a potentially beneficial effect on health when consumed as part of a varied diet on a regular basis.

Nutrigenomic Education and Professional Training

The ability of nutrigenomics to provide options for individual health management, has sparked a new era in nutrition and dietary practice. However, the incorporation of nutrigenomics into education and training of professionals has been slow due to a) time lag involved in establishing incidence and its clinical validity of a given genetic mutation in large number of subjects; b) time to develop affordable diagnostic tests and c) ability of the physician and clinical facility to offer the test and d) availability of appropriate supplements or functional foods that can truly benefit the subject [14]. Also, there are limited numbers of large scale population studies documenting the nutrigenomic concepts [15-17]. Furthermore, the understanding of nutrigenomics needs a substantial background in genetics , biochemistry and some basic understanding of clinical chemistry principles. Some course formats to introduce nutrigenomics definitions and concepts into dietetic curriculum are discussed in articles by Prasad et.al. [18] and Fenech et.al [19]. Also, there is a urgent need for more emphasis on to establish guidelines for proficiency testing, certification and ethical aspects related to the operation of nutrigenomic laboratories, especially those offering tests directly to the public [20]. Some of the key immediate goals to meet the educational challenges of nutrigenomics for this decade are listed below:

1) To increase awareness and understanding of nutrigenomic concepts and testing among primary care physicians, internists, pediatricians and physicians involved in clinics devoted to chronic diseases.
2) To increase more in-depth training of genomic, proteomic and metabolomic concepts to dietitians, nutritionists and genetic counselors.
3) To introduce racial, cultural and ethnic aspects of food into curriculum of students in dietetics, nutritional sciences and genetic counseling.
4) To train Food and Nutritional professionals work with food industry, government, allied health professionals and media to ensure that the public has accurate information on nutrigenomics and enforce consumer protection issues. This will also be very important in dealing with dermagenetics that focuses on genetic tests related to skin health and advocate nutritional and topical products containing cosmeceuticals [21].

It is hoped that further research and careful attention to the above discussed issues will greatly contribute towards the application of nutrigenomics in personalized medicine and help towards reducing the health care costs associated with chronic diseases.

References

[1] Gurvitz D, Bregen-Eschet Y. Personal genomic services: whose genomes?. *Eur. J. Clin. Genet.* 2009;17: 883-889.

[2] Saukko PM, Reed M, Britten N, Hogarth S. Negotiating the boundary between medicine and consumer culture: online marketing of nutrigenetics tests. *Soc. Sci. Med.* 2010; 70:744-753.

[3] Cervinski MA, Tsongalis GJ. Direct-to-Consumer genotyping: are we ready for a brave new world?. *Clin. Chem.* 2010;56:1056-1060.

[4] Ozemir V, Godard B. Evidence based management of nutrigenomic expectations and ELSIs. *Pharmacogenomics* 2007;8:1051-1062.

[5] Levesque L, Ozdemir V, Gremmen B, Godard B. Integrating anticipated nutrigenomics bioscience applications with ethical aspects. *Omics* 2008;12:1-16.

[6] Ferguson L. Nutrigenomics approaches to functional foods. *J. Amer. Diet Assoc.* 2009; 109:452-458.

[7] Ghosh D. Personalized food: how personal is it?. *Genes Nutr.* 2010;5:51-53.

[8] Thompson GR, Grundy SM: History and development of plant sterol and stanol esters for cholesterol lowering properties. *Amer. J. Cardiol.* 2005; 96 (suppl 1):3-9.9.

[9] Plat J, Mensink RP. Plant stanol and sterol esters in the control of blood cholesterol levels. 2005; 96(1A): 15D-22D.

[10] Weggemans RM, Zock PL, Tai ES, Ordovas JM, Molhuisen HO, Katan MB: ATP binding cassette G5 C1950G polymorphism may affect blood cholesterol concentrations in humans. *Clin. Genet.* 2002; 62, 226-229.

[11] Herron KL, McCrane MM, Waters D, Lofgren IE, Clerk RM, Ordovas JM, Fernandez ML: The ABCG5 polymorphism contributes to individual responses to dietary cholesterol and carotenoids in eggs. *J. Nutr.* 2006; 136,1161-1165.

[12] Lupton JR. Scientific substantiation of claims in the USA: focus on functional foods.Eur.J.Nutr. 2009;48(Suppl 1):s27-31.

[13] Aggett P. The process of assessment of scientific support for claims on food. *Eur. J. Nutr.* 2009; 48 (Suppl 1): s23-26.

[14] Subbiah MTR. Nutrigenetics and nutraceuticals: the next wave riding on personalized medicine. *Transl. Res.*2007;149:55-61.

[15] Tierney AC, Mcmonagle J, Shaw DI, Gulseth HL, Saris WH, Paniaqua JA, Golabek-Lesszczynska I et.al. Effects of dietary fat modification on insulin sensitivity and other risk factors of the metabolic syndrome-LIPOGENE: a European randomized dietary intervention study. *In. J. Obes.* (Lond). 2010; 35:800-809.

[16] Lairon D, Defoort C, Martin JC, Amlot-Carlin MJ, Gastaldi M, Planells R. Nutrigenetics: links between genetic background and response to Mediterranean-type diets. *Public Health Nutr.* 2009; 12:1601-1606.

[17] Konstantinidou V, Covas MI, Munoz-Aguayo D, Khymenets O, de la Torre R, Saez G, Tormos Mdel C et.al. In vivo nutrigenomic effects of virgin olive oil polyphenols within the frame of the Mediterranean diet: a randomized controlled trial. *FASEB J.* 2010; 24:2546-2557.

[18] Prasad C, Imrhan A, Rew M. Introducing nutritional genomic teaching in undergraduate dietetic curriculum. *J. Nutrigenet. Nutrigenomics.* 2011; 4:165-172.

[19] Fenech M, El-Sohemy A, Cahill L, Ferguson LR, French TA, et.al. Nutrigenetics and nutrigenomics: view points on the current status and applications in nutrition research and practice. *J. Nutrigenet. Nutrigenomics.* 2011; 4:69-89.

[20] Castle D, Reis NM. Ethical, legal and social issues in nutrigenomics: the challenges of regulating service delivery and building health professional capacity. *Mut. Res.* 2007; 622:338-343.

[21] Subbiah MTR. Application of nutrigenomics in skin health: Nutraceutical or cosmeceutical? *J. Clin. Aesth. Dermatol.* 2010; 3:44-46.

Index

A

B

C

D

F

G

H

I

N

O

P

Q

R

S

T

U

V

W

X

Y

Z